A QUALITY REVOLUTION IN MANUFACTURING

A QUALITY REVOLUTION IN MANUFACTURING

Victor R. Dingus and William A. Golomski

**Published by
Industrial Engineering and Management Press
Institute of Industrial Engineers**

**Library of Congress
Cataloging-In-Publication Data**

A Quality revolution in manufacturing/edited by
  William A. Golomski, Victor R. Dingus.
  p. cm.
  ISBN 0-89806-075-3 : $49.95 ($30.95 to IIE members)
  1. Quality control—Case studies.
  2. Production management-Quality control—Case studies.
I. Golomski, William A., 1924- II. Dingus, Victor R., 1948-
TS156.Q365   1988   658.5'62—dc19   88-31843
                                                         CIP

Industrial Engineering and Management Press, Norcross, Georgia 30092

© 1988 by the Institute of Industrial Engineers. All Rights Reserved. No part of this book may be reproduced in any form without written permission from the publisher. The views and concepts presented are those of the authors. Publication by the Institute of Industrial Engineers does not in any way constitute endorsement or approval of the book's contents.

Published in 1988.
Printed in the United States of America
93   5 4 3
ISBN 0-89806-075-3

Additional copies may be obtained by contacting:
Publication Sales
Institute of Industrial Engineers
25 Technology Park/Atlanta
P. O. Box 6150
Norcross, Georgia 30091-6150

TABLE OF CONTENTS

Part I. Top Management Perspective

Chapter 1. A Top Management Quality Revolution 1
William R. Garwood and David M. Sandidge

Chapter 2. One Step at a Time: Implementing a Quality Process . . 11
Lewis Lehr

Part II. Industry Applications

Chapter 3. Total Quality Issues and Activities in the
Defense Industry 19
Hugh Jordan Harrington and Jack B. ReVelle

Chapter 4. Ford Quality Is Job Number One 39
Dean E. Smith, Jr.

Chapter 5. Managing for Quality Improvement in the
Eastman Chemicals Division 49
Paul Hammes, John Wallace, and Lee McConnell

Chapter 6. Xerox Process Qualification for Suppliers 67
Martin J. Madigan

Chapter 7. Quality Improvement in the Pharmaceutical Industry . . 77
Thomas L. Fine

Chapter 8. Statistical Process Control Application to the
Aluminum Extrusion and Drawn Tube Process 87
Robert F. Wolf

Part III. Management Tools

Chapter 9. Application of Quality Cost Concepts 103
William A. Golomski

Chapter 10. Industrial Engineer Meet Dr. Deming: A Matter
of Corporate Survival 117
Victor R. Dingus

Part IV. A Vision of the Future

Chapter 11. The Quality Imperative in the New Economic Era . . . 133
Myron Tribus and Yoshikazu Tsuda

Preface

Recently, the American marketplace has been faced with an emphasis on competing with Japanese and European manufacturers. There is a growing awareness in the United States of the lack of quality of American-made products when compared to some foreign-made products. American companies have begun to realize that it is not enough to manufacture products which incorporate the latest technology. Products must be up-to-date to attract buyers, but more than that the products must be able to meet and withstand the customer's expectations. The products must be durable, must serve their purposes well, and when there is a problem, manufacturers must solve the problem on the first try.

This collection of invited papers represents a cross-section of leading approaches to quality improvement in U.S. manufacturing industries. The authors are senior executives and other recognized professionals in the field.

The book is divided into four sections beginning with the most important aspect of implementing a quality improvement program: top management support. Chapters 1 and 2 detail the importance of obtaining top management support and leadership. In order for the process to be successful management must get involved in the day-to-day operations, and be trained in the tools of quality management. Chapters 3 through 8 are case studies of companies who have and continue to strive for excellence in manufacturing through quality management techniques such as defect prevention, statistical process control, teamwork concepts, and performance management. Presented are the trials, tribulations, and successes experienced by companies in industries such as: aerospace, automotive, chemicals, pharmaceuticals, metals, and office equipment/computers. Chapter 9 introduces quality cost systems to be used as an aid in prioritizing quality improvement projects, studying cost trends to re-allocate resources, focusing multistep operations, measuring performance, and balancing efforts in reducing variation in design versus manufacturing. Chapter 10 is a discussion of the importance of applying Deming's 14 Points of Management to industrial engineering functions within the manufacturing process. And finally, Chapter 11 takes a look at the future and the imperative of a quality improvement mindset, process, and gameplan.

The American corporations who take an inward look at themselves and adapt a quality improvement process in their manufacturing arenas will not only be one step closer to achieving quality excellence, they will be narrowing the competitive gap between foreign- and American-made products. These companies will also be able to hold on to, if not increase, their share of the American marketplace. The companies who have not realized the importance of keeping up with the Joneses will not only lose ground internationally, but will be edged out by their American competitors.

Victor R. Dingus
William A. Golomski

Part I. Top Management Perspective

Chapter One
A Top Management Quality Revolution

William R. Garwood
Executive Vice President and Works Manager
David M. Sandidge
Tennessee Eastman Company

INTRODUCTION

Foreign competition has acquired a large share of business, earnings, and jobs from some well-known United States companies. This deindustrialization of America will continue unless management acts to reverse the trend—many people are unaware of how pervasive the problem is. Figure 1 illustrates the domestic market share imports have captured in a wide variety of industries. Even with considerable attention to the trade imbalance, the imports' share of the market has increased in the last four years. One of the primary reasons for this market penetration is superior product quality and value.

If such losses have not already occurred in your industry, they will in the future. Imagine being the chief executive officer of a company and waking up one morning to find a competitor producing a better product at a lower price than yours. This is precisely how the competitor is gaining market share. Thus, you face an undesirable range of choices for the short term:

1. cutting price;
2. sorting out bad products and increasing your costs;
3. giving up your customers and reducing the scope of your operations.

All of these are short-term decisions, none of which will keep you in business very long. The only long-term alternative is to change the way you operate your business. Eastman knows this from experience. The competitive quality edge of one of its products was lost in the late 1970s. After Eastman realized a problem with the product existed, the company allocated additional resources to improve the product, brought teams together from different functional groups, and began to work closely with customers.

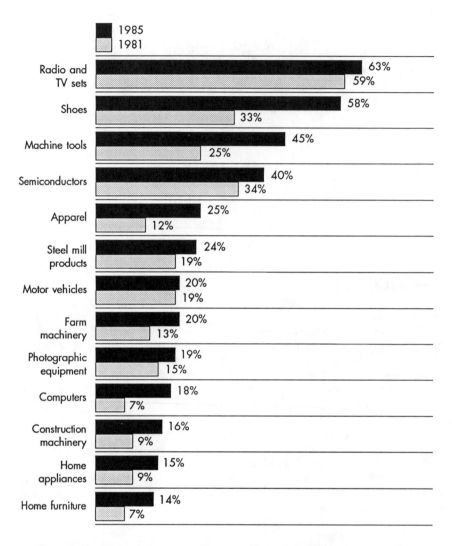

Figure 1. Imports as a Share of Domestic Consumption

Reprinted by Permission of *The Wall Street Journal*. Copyright Dow Jones & Company, Inc. New York City. 1986. All Rights Reserved.

After quality improvements over several years (which cost more than $10 million), the company recaptured the business, reduced costs, and regained its position of quality leadership. Through this experience Eastman proved that improved quality can lead to lower unit costs and a stronger competitive position. Revolutionary thinking is essential to developing the sense of urgency necessary for competing in the new economic age. The new global economic situation will not allow a return to the era when market share could be protected from high-quality competition.

A Sense of Urgency
A company must continually improve product and process quality for long-term survival and prosperity. Companies cannot wait until they've lost business to begin to change their operations. If they do, they will not survive. Because of the scope of the change being implemented, there is a time lag from one to five years after management adopts a new way of thinking before results are realized. During this time competitors have a tremendous advantage and may widen that advantage by continuing to improve their operations.

The comparative advantage extends to all aspects of operations. A comparison of the cost advantages of an operation managed for quality is presented in Table 1 (Harbour and Associates 1982).

The new management system has improved all aspects of operations for a variety of companies. General Electric slashed the total manufacturing time for dishwashers from three days to less than eight hours (Sease 1986). A 20% increase in unit productivity occurred at Dana Corporation (Productivity 1986). Inventories have been reduced by 40-90% at Omark (Walters 1984). Harley Davidson has increased its percentage of defect-free finished products from 50% to more than 99% (Willis 1986). Tennessee Eastman Company has increased annual sales by more than $20 million due to its quality efforts. Companies that have discovered the power of quality management are seeing dramatic results.

Table 1
Cost Advantage of Japanese Auto Manufacturers versus U.S. Manufacturers

Category	Cost Advantage, $/Car
Labor (60 vs. 120 hours)	550
Advanced Technology (greater use of robots)	73
Quality Control (more inspectors, repair men)	329
Excess Inventory (work in process)	550
Materials Handling (extra manpower)	41
Better Use of Labor (fewer job classifications)	478
Absenteeism	81
Assembly Line Relief System (shutdown vs. "tag")	89
Cost of Union Representation	12
Total Cost Advantage	2203
Shipping, Handling, Import Duties	-485
Total Cost Advantage	1718

Revolutionary Principles

The quality revolution is based on two principles: continual quality improvement, in all aspects of the business, and defect prevention. Companies that adopt these principles of continual improvement provide a constancy of purpose for individuals throughout any diverse organization —the underlying assumption is that the current performance level can be improved.

Many companies agree with, but few practice, the principle of defect prevention. Preventing defects is more cost effective than inspection and sorting. The prevention strategy has several advantages over the traditionally used detection strategy:

1. Feedback from control points in the process is more timely and traceable than results from mass inspection of final products.
2. Statistical methods are used to determine when problems are present instead of comparing products against some (often) arbitrary specification.
3. Inspection processes are not perfect. Therefore, some defective products get to the customer when the detection strategy is used. The prevention strategy stops nonconforming products from being produced.
4. Control of process variables and raw materials will prevent many problems from occurring in the first place.

The New Way of Managing

The revolution in management thinking that will enable you to compete through quality of processes, products, and services is conceptually straightforward. The difficulty comes in believing and executing changes. Although there is no one detailed model for implementing the changes in your company, there are several key elements and characteristics of the new way of managing.

Importance of quality. Quality is as important a measure of a company's effectiveness in satisfying customer needs as earnings and stock prices are measures of the effectiveness of financial performance. The ultimate goal of the corporation is not short-term profits. The goal is to survive, grow, and prosper—that depends on satisfying customer needs. Therefore, improving the quality of products and processes is of paramount importance to long term survival.

Involving everyone in improvement. Managers have asked employees to bring their bodies in the plant but leave their brains behind for too long. As a result, a self-fulfilling prophecy has developed. People who operate the process have never been asked for their ideas and, therefore, feel they have nothing to contribute—they just work the job. The new management style requires that we allow employees to develop more of an ownership of the processes they work in, instead of just working the job. The ownership attitude leads to working and improving instead of just working. As a result, the most valuable resource, people, contribute to the company with their problem-solving talents.

Focusing on the process. All products come from a process. This applies whether the product of the organization is a widget, a chemical, a retail sale, a consulting service, a computer software package, or an engineering design. Process (or system) improvements can only be made in two ways:

1. removing special causes of variation;
2. improving the system.

The new management style defines management's job in the following way (Tribus and Tsuda 1985):

1. Management works on the system; workers work within the system.
2. Management improves the system with the workers' help.

An obvious short-term solution is, when problems arise, to bring all hands to work on the system. Blaming people is no longer the answer.

Understanding and dealing with variation. The output from any system exhibits variation. Statistics is the tool used to interpret and deal with variation. Without a basic knowledge of statistics at all levels in the organization, process operators cannot identify and distinguish between random variation and variation resulting from a special cause. Using statistical tools properly is of great importance in identifying problems, eliminating them, and improving the system. Because of little exposure to statistics in educational curricula, most people have very limited knowledge of variation and the tools to control it. This is particularly true in engineering disciplines. Engineers are taught, to a large extent, that people live in a deterministic world. The truth is otherwise.

Emphasis on customers. Quality is defined by customers. You cannot identify and meet your customers' needs without getting close and working with them. A good measure of whether you are meeting your customers' needs is repeat business. Tom Peters defines it this way: "The be-all and the end-all of business . . . is repeat business — which only comes as a result of long-term customer satisfaction" (Peters and Austin 1985). Customers are interested in two kinds of quality: the quality of design and the quality of conformance. That is why the quality revolution must apply not only to producing uniform products, but also to improving the quality of the design of those products.

Managers often confuse quality with features. Multiple features do not necessarily constitute a high degree of quality. For example, compare the products of home-building contractors. A house with a built-in intercom, three-car garage, skylights, and a jacuzzi has many features. But these features have no relation to the grade of lumber used, straightness of lines and framing, and quality of carpentry and other workmanship. A company can define the features needed by its customers. In fact, many times customers are unaware of the features they want and would be willing to pay for. But quality of the product is always defined by the customer.

Building an infrastructure to support the effort. Management must provide an infrastructure of training, management systems, and teamwork to support the new way of managing. In particular, employee involvement in problem solving may be very difficult to start. It is impossible to reverse 20 years of noninvolvement overnight. Training in basic problem-solving techniques such as cause and effect diagrams, pareto analysis, brainstorming, group dynamics, and team-meeting leadership skills must be completed. Management systems must also support and reward the emphasis on quality of product and process and the use of a team approach to problem solving. Behavior is a function of consequences—people will not change their behaviors unless there are systems in place to reinforce and support them. A strategy must exist for involving everyone in improvement. A few teams of volunteers may be sufficient to start the effort. Full benefits, however, will not be realized until everyone is involved.

What Should We Expect To Happen?
After World War II, industry in Japan took 20 years to transform. American business simply cannot take that long, and it really shouldn't need that much time. Companies can learn from each other, from the Japanese, and from the many fine domestic firms that have adopted this new management style. Industry must learn what they have done, and adapt it, and move on much more quickly.

As this transformation is made, companies will see some definite stages. Initially, there will be a few believers in the company. They, in turn, will create a few converts, and there will be a few successes. Word will spread. At this stage there will be confusion because many people will want to act before they know how. At that point a crossroad is reached. If employees are educated in the theory and use of the tools, as well as provided an infrastructure of support, a grass-roots explosion will occur, and success upon success will be realized. If, on the other hand, leadership and management infrastructure are not provided, no amount of exhortation will keep the effort from dying.

Costs may rise initially because of the need to continue detecting and sorting product, and training employees. But as the principles of defect prevention and continual improvement are applied to processes, improvements will come quickly, and the costs of nonquality will decrease. Less waste in material, time, and money will result. More accurate information will be available from the process, thus enabling further breakthroughs in performance to occur.

The transformation will also have a profound impact on an organization's people. Increased information sharing and more open communications within and among departments will take place. People who are involved in team problem-solving efforts will experience higher morale. Your company may experience some difficulty in getting middle management involved—they are truly caught in the middle. Top management is supportive and leads the effort, and the process operators want to get involved and contribute. The middle manager has to handle these pressures as well as lead the organization forward. Consequently, the middle of the organization will be getting flatter. Understandably, these changes could make the middle manager feel threatened and unsure about the future. An important part of the infrastructure discussed earlier is setting up supporting management systems to help and support middle management in implementing the new management style. Developing a vision of what your company will be like when the quality transformation is fully achieved will also help guide your efforts. An example of such a vision is presented in Figure 2.

Keys to the Revolution
Although there are many important elements of the new management style, three points deserve additional comment. Transformation is difficult to begin, and the new management style will be virtually impossible to institutionalize without these three elements:

1. *Dedication by top management.* Involvement and leadership is emphasized not just support. Your management team should learn the theory and gather knowledge to apply the concepts. Deming says, in effect, doing their best is simply not good enough. They must know what to do (Deming 1982). There are many consultants in the field who concentrate on a particular aspect of quality. Managers should learn from several consultants and then develop an overall strategy they can apply in their businesses based on fundamental concepts. Managers must lead the effort if it is to succeed.

---

1. Open communications among all employees
2. Teamwork at all levels
3. Full trust and cooperation among all organizational units
4. Process operations under statistical control
5. Low levels of defects
6. High employee morale
7. Plant at capacity (sold out)
8. Customers actively seek the company as suppliers
9. First in the market with profitable new products
10. High growth
11. Lower factory and overhead costs and increased earnings
12. Meetings held for exceptions only
13. Plant designs are balanced for quality, process stability, reliability, ease of operation, and costs
14. Appraisal/reward system is accepted as fair by all
15. Company looked at as a model business by the community and government
16. Everyone wants to work for the company

Figure 2. Vision of a Manufacturing Operation Managed for Quality

2. *Quality improvement planning.* Without proper planning to focus efforts, maximum use of our resources is not achieved. This planning must be continuous and dynamic, always responsive to changing customer needs.

    Improvement comes on a project-by-project basis. An understanding of the improvement cycle of control and breakthrough, illustrated in Figure 3, is essential.

    The manager's role in control is to ensure that systems exist for identifying out-of-control situations and eliminating the causes of these upsets. The manager's role during breakthrough is to supply the means to determine if a new level is needed and to execute the problem-solving steps required to reach the new level (Juran 1964). Planning control and breakthrough projects will ensure that you are "doing the right things."
3. *Quality training and execution.* New skills everyone will need include problem solving, teamwork, use of statistical methods, and group dynamics. This training is a prerequisite for getting everyone involved.

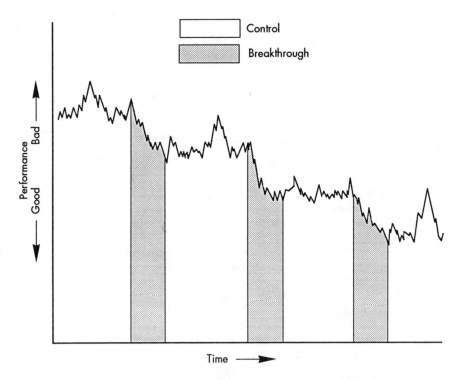

Figure 3. Interrelation of Breakthrough and Control

From J. M. Juran, *Managerial Breakthrough*. McGraw-Hill, New York, 1964. Used with permission.

You cannot accomplish something new if you don't know how to begin. Training followed by a period of inaction will undermine what is learned. The infrastructure should support team formation as soon as possible after training to get people using and becoming comfortable with their new skills. Training, combined with management systems for execution, will help us "do things right."

The crisis facing much of industry is real. The changing economic climate will result in intensified competition. Nothing less than a revolutionary change in management thinking about quality will enable companies to survive and prosper. Adopting the philosophy of quality management is a company's best way to ensure improved quality, improved productivity, lower unit costs, increased market share, growth, and prosperity. Good luck.

References

Deming, W. Edwards. 1982. *Quality Productivity and Competitive Position.* (Published by MIT, Center for Advanced Engineering Study, Massachusetts Institute of Technology. Cambridge, MA: 13.

Harbour and Associates. September 1982. Quality day. Presented at Jackson Community College, Jackson, Michigan.

Juran, J. M. 1964. *Managerial Breakthrough.* McGraw-Hill. New York: 6.

Peters, Thomas J. and Austin Nancy K. 1985. *A Passion for Excellence.* Random House, Inc. New York: 77

Productivity, Inc. 1986. *Productivity.* Stamford, Connecticut. Pamphlet 3, vol. 6.

Sease, Douglas R. 16 September 1986. How U.S. companies devise ways to meet the challenge from Japan. *Wall Street Journal.* pp. 1, 25.

Tribus, Myron and Tsuda, Yoshikazu. 1985. *The Quality Imperative in the New Economic Era,* MIT Video Course, Center for Advanced Engineering Study. Massachusetts Institute of Technology, Cambridge, MA.

Walters, Graig. March 1984. Why everybody's talking about just-in-time. *Inc.,* pp. 77-90.

Willis, Rod. March 1986. Harley Davidson comes roaring back. *Management Review.* pp. 20-27.

Chapter Two

One Step at a Time: Implementing a Quality Process

Lewis Lehr
Retired-Chief Executive Officer
3M Company

Implementing a quality improvement process at 3M has been, and continues to be, some of the hardest work the company has undertaken during its 85-year history. Not only has 3M instilled a new corporate quality philosophy during the last six years, it has built a concrete, step-by-step process for conforming to customer expectations throughout its worldwide organization.

3M has devoted millions of hours and significant resources to developing a quality process that fits its company and its needs—and which is yielding important cost, profit, and market improvement results.

Improvement did not come overnight, nor without a great deal of learning. What began as a hopeful new "program" for 3M has evolved into a deeply ingrained process to which the company is totally committed—and which will require constant improvement and fine-tuning to keep pace with the competitive realities of its markets and, most important, the constantly changing expectations of 3M's customers.

Along the way to 3M's current level of quality enlightenment and results, the company has learned some important lessons about how—and how not—to go about installing a quality improvement process. A resume of the 3M experience may provide valuable insight to help other companies launch improvement processes.

Competitive Advantage

3M took a close look at a quality improvement process for the same reasons virtually every company does—competitive pressures and the need to improve quality at less cost. Although high-quality, innovative products and services have been 3M hallmarks almost since the beginning of the company, by the late 1970s some divisions were being threatened by increased competition.

Recent history in several markets that 3M pioneered lent particular urgency to the company's efforts. For example, the market for audio cassette tapes, a product which 3M invented, had been eroded by competition, and the market for copying products, which 3M also developed, was slipping from 3M's grasp. It was obvious that the company's formidable technical and new product development process could no longer guarantee continued success. A return to market leadership was a compelling reason to move vigorously ahead with quality improvement.

To meet this tougher competition, 3M needed a fundamental change in its traditional quality process. The challenge was to undergo an internal metamorphosis in quality management to respond effectively to the external forces of competition and market expectations.

Great Expectations
The change began in 1979 and 1980 with a wave of management awareness and enthusiasm for the quality philosophies and techniques of a handful of established—as well as newly emerging—quality "gurus." The various, sometimes overlapping, management approaches of these quality philosophers, some of whom borrowed heavily from the Japanese experience (and others of whom, ironically, had helped the Japanese lay the foundation of their quality strategies), appeared to fit the situation and needs of a number of 3M's divisions.

After examining the best ideas, the corporation embraced a quality approach founded on error prevention and total organizational involvement in the quality process. 3M employees attended seminars that helped the employees understand the new quality philosphy of error prevention. 3M believed a track to run on for improving products' quality and strengthening competitive position had been found.

"Quality" quickly became a new buzzword throughout the company. Committees and steering groups were appointed to implement the new quality philosophy and to help bring quality out of the factory and into sales and marketing, laboratory, service, and distribution settings. The company felt that great things were about to happen with quality at 3M. They did happen—but not in the way 3M originally envisioned.

Reality Hits Home
3M's great expectations for a new quality process that would return immediate and measurable results were quickly tempered by the reality of two facts. First, the organization discovered that a quality process must be adapted to, not adopted by, an organization.

Although some consultants may want people to believe otherwise, a quality process is not an off-the-shelf item. That is, one specific process approach or another, no matter how seemingly appropriate or logical, cannot be installed intact into an organization and made to perform in day-to-day operations the way it does on paper—or even for another organization.

Every company, particularly one such as 3M, with a long history of success and a carefully constructed culture has valid systems and procedures in place that should not and cannot be abandoned in favor of a single improvement prescription. Rather, a new process must be carefully studied and judiciously adapted when and where it makes the most sense for productive improvement.

Next, 3M discovered that quality improvement can only be advanced by a slow, painstaking change. The company must become outwardly directed (toward the customer), rather than inwardly focused. Long-range thinking and planning must replace the penchant for short-term results. Authoritarian styles of management must be replaced by styles favoring the productive participation of all levels of employees in the improvement process. Communication channels, perhaps the single most important aspect of improvement, must be shortened and streamlined.

These are far-reaching changes that, as with the aging of a fine wine, cannot be hurried. They take time, care, and prudent planning. People don't change overnight. And, as the quality process must realistically be based on people, neither will cultural change occur in the short order as people would wish.

These realizations did not occur to us all at once, however, but were apparent only through long trial and error in developing what has become the 3M-quality-improvement process.

Initially, 3M established the staff quality group, which became responsible for educating top management and conducting awareness training for midlevel management.

[*A short digression:* The vertical organization of 3M, with approximately 50 autonomously operating divisions plus tens of foreign subsidiaries, each responsible for sales, growth, and profit targets, dictates that the corporation leads more than directs major new management initiatives. The history of the new quality process has been marked by gradual, rather than precipitative, management involvement.]

Start-up training involved management orientation to the new error-prevention philosophy and a step-by-step general outline for implementing of an error-prevention and corrective-action process. Management went back to their operating units with a lot of enthusiasm and a game plan for making quality improvement happen.

The leap from the textbook to the trenches, however, was more than the initial process could bear. Increasingly, the company discovered that not only were people unsure of how to install an error-prevention process into their organization, but the process as originally conceived was limiting for the business situations within 3M's divisions.

Also, the improvement effort was sometimes misinterpreted as a motivation program, a productivity program, a cost-reduction program, a resource reduction effort or just another fad. Some lack of total management commitment within divisions and subsidiaries also hindered early progress. It took time for top management to fully comprehend the magnitude of quality improvement and to understand their obligation to lead the process (a step that is still occurring within some divisions).

After a honeymoon period marked by enthusiasm and a great deal of training and communication, employees found themselves on a plateau. The company asked: "Where does 3M go from here?" The answer was to push beyond the first process model and develop an improvement system that fit 3M's specific culture and needs. When the company realized that meaningful improvement depended on designing its own quality process, the effort began to pick up steam.

Thanks to corporate commitment to quality improvement, the tenacity of the staff quality group, and the leadership of several divisions, the improvement process evolved from a "program" to a process as 3M gained new input from outside consultants and gained more insight through hands-on experience.

Gradually, 3M came to understand that quality improvement is a precise science, not an art, and involves many different management and technical disciplines.

Some people originally believed that waving a magic wand over the organization would suffice to get the process started and keep it going. The company learned instead that to tailor the quality process, a tremendous amount of knowledge and training in many areas, ranging from quality circles and statistical process control to team building and performance management techniques, is necessary. To generate measurable quality improvement—and to perpetuate the process—people must be equipped with a wide range of new skills.

Lessons Learned

The 3M quality improvement experience over the past six years has resulted in a quality process comprised of a strong, real-world philosophy made substantive by a pattern of consistent, disciplined action. In other words the company had a concept for building quality improvement into its operations and a blueprint for that concept to ensure that the building of quality improvement is constructed step by step.

3M began quality improvement by making several assertions about quality that to traditionalists (that is, those who still believe quality is a "factory problem") may seem heretical, but which are absolutely essential to the foundation of a quality process.

These assertions were as follows:

1. Quality improvement at 3M is a worldwide process for the company involving all people, disciplines, operating groups, and staff groups.
2. Quality improvement involves short-term and long-term improvement of personal performance, products, and services.
3. Quality improvement emphasizes an open organizational structure—that is, one which is customer-driven and involves improving relationships among people to improve performance results.
4. Top management commitment and leadership is essential for improvement. Management must lead the process downward in the organization.
5. The quality process cannot be superficial to the organizations daily life, but must be deeply ingrained in issues relating to cultural change and growth—it is a way to manage.
6. A system for training management to lead the process, and for implementing of the process with planned follow-up in each

operating unit, must be established. [At 3M, this feature has been spearheaded by the staff quality group, along with other staff support groups who provide training and consultative services to 3M's divisions, subsidiaries, and staff groups.]
7. The process is not static, but will continuously improve with experience and new input.
8. The process is not comprised of a single defined improvement approach, but rather it incorporates several approaches that have been adapted and refined to fit the company.

These approaches can be graphically illustrated as follows:

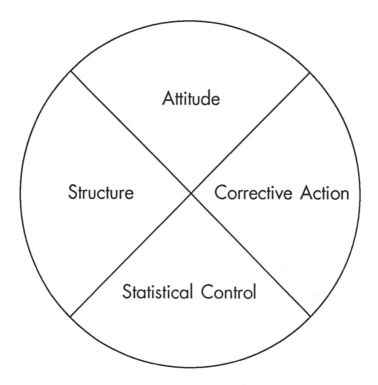

Figure 1. 3M Quality Improvement Process

The mixture of these various approaches is what 3M terms "total quality": the process of continuous organizational improvement to better enable 3M to conform to its customers' expectations.

To support these assertions or principles, 3M has developed a philosophy of quality management. This philosophy states in very brief, unequivocal terms the approach management must take in defining, measuring, quantifying, standardizing, and leading the quality process. 3M's philosophy is a concise summation, born through many years and many instances of both success and failure, of how management must think and behave to succeed in the permanent pursuit of quality improvement.

How important is a feasible quality philosophy? Experience has taught 3M that it is the single most important component of the improvement process. Without a quality philosophy, nothing will happen—the quality process will flounder for lack of real conviction and the single-minded purpose necessary to implement genuine change and a permanent lifestyle of continuous organizational improvement.

The company recognizes that the management who make the quality philosophy their way of life have the greatest chance of success in the improvement process. Conversely, 3M also recognizes that the management who fail to embrace the philosophy stand the most chance of failure.

But if the first lesson of quality improvement is establishing a clear quality philosophy, the second lesson is that the philosophy must be implemented through action.

To implement 3M's quality philosophy, management was equipped with the understanding and the support necessary for a never-ending series of short-term and long-term action steps that place the improvement process in motion and keep it going—forever.

These action steps are the real blueprint for the improvement process. The steps describe how management must commit, plan, organize, educate, and communicate for quality improvement and how sensitivity to customer expectations and quantification of quality performance through the organization are achieved.

At 3M, the quality philosophy and action steps together make quality improvement a permanent way of life and a key part of its overall business and market strategies.

A Beginning, But Never An Ending

After six years of hard work and concerted effort, there are still many opportunities for quality improvement within 3M. Different areas of the company have achieved varying levels of commitment and proficiency with the process. Some departments are making outstanding progress, some are making adequate progress, and others still need to get started.

But one thing is certain: There is no going back to a traditional quality approach. 3M's quality process will never end, only improve and move ahead.

3M sees its process as a "living laboratory" in which each unit's experience with the improvement process adds to its understanding and knowledge of how the company can improve the process throughout the company. In addition, the company has begun working with other companies to share with them the benefits of 3M's experience and insights, an undertaking which should help 3M to understand its own improvement process even better.

3M has learned much about quality improvement since the company took its first steps. 3M recognizes that there is much still to learn. And learn 3M will—one step at a time.

Reprinted by permission of and copyright 3M Company 1986.

Part II. Industry Applications

Chapter Three
Total Quality Issues and Activities In the Defense Industry

Hugh Jordan Harrington
Senior Research Specialist for Human Resources
Research and Development
Jack B. ReVelle
Chief Statistician, Quality Management Staff
Hughes Aircraft Company

Before discussing some of the current quality improvement issues and activities in the defense industry, it is appropriate to define how the defense industry operates—in particular, the characteristics that make this industry unique in relation to all other commercial industries. Both recent and emerging changes in business operations in this industry will also be identified. This background information should make some of the results of the Hughes'-conducted total quality survey more easily understood. The survey results constitute the main thrust of this chapter.

Uniqueness of the Defense Industry
Sole Customer. The defense or aerospace industry is unique among industries primarily because it has a single customer controlling the market force—the U.S. Department of Defense (DOD). This customer, unlike any other, can determine for all sellers the terms and conditions under which it will purchase goods. There are constraints on prime contractors and their subcontractors with respect to schedule, budget, technical requirements, manufacturing methodologies, and specifications. Moreover, the customer is an integral part of all design decisions. The customer has total visibility and access to every aspect and record of the entire operation and regularly conducts systemwide audits, over and above the daily in-plant surveillance that exists.

For illustration, "the customer" has been represented as a single entity. In fact, there are multiple layers of "the customer," starting with Congress and its special committees, the Government Accounting Office, the Department of Defense, and then each of the services (Army, Navy, Air Force, Marines), and within each of the services, its particular arm—for example, the Program Management Office, Contract Administration Office, and Plant Representative Office. This level of complexity further constrains defense contrators.

Accountability. The funds for defense make the aerospace industry almost entirely dependent on the public, and, therefore, not only subject to governmental regulations and scrutiny, but also to the public's and the media's attention. No one denies the public and the government the right and obligation of such attention. Events in recent years involving waste, fraud, and abuse, however, have been so well publicized that a justifiable concern has perhaps become an unreasonable degree of suspicion and mistrust. The level of punitive actions has dramatically increased in the form of suspended progress payments and debarred contractors, preventing them from bidding on any contracts until they are exonerated.

The government and the public's fear of large, continuing deficits reflect a legitimate concern for fiscal realities. As a consequence, the greatest attention has been focused on defense spending, which accounts for one third of the federal budget. Yet, only one third of that budget represents prime contract awards for such hard goods as aircraft, electronics and communication equipment, missiles, and space systems. Moreover, less than 10% of these contract awards represent pretax profit. Even the percentage of profit a contractor is allowed to earn is limited.

Sophisticated specialization. A third characteristic which makes the defense industry unique is that it has become highly sophisticated and specialized. New high-technology product lines include the Strategic Defense Initiative (SDI), Very High Speed Integrated Circuits (VHSIC), and high-speed supercomputers used for aerodynamic modeling and cryptography. Particle-beam weapons, electronic counter measure (ECM) aircraft and other electronic weapons systems (EWS), and new and advanced composite materials that reduce weight and save fuel are also included among new high-tech products. "Pushing the state of the art" on each new contract makes it more difficult to focus on quality improvement, as might be done for a "tried and true" product, manufactured by the thousands year after year.

The Aerospace Total Quality Survey

To answer the question, "What quality-related activities are currently underway in the defense industry?" We rely primarily on a survey conducted. The survey consisted of a major section on quality improvement issues and activities and a second section on statistical quality and process control. This chapter will present the results of the survey in that order.

The survey was distributed to 120 aerospace and defense-electronics companies (and some of their major divisions) throughout the nation. It was directed by name to the senior quality vice-president and director of each company.

A total of 90 surveys (75% response rate) were returned in time for analysis. Half of the top 50 defense contractors in the nation (listed in *Aviation Week and Space Technology*, May 12, 1986) participated. These participants alone account for more than 40% of the defense contracts awarded by the DOD. Table 1 presents a partial list of participating defense contractors.

Table 1
Partial List of Participating Companies

Allied Signal	Lockheed
Eastman Kodak	Martin Marietta
Emerson Electric	McDonnell Douglas
Fairchild Industries	Morton Thiokol
Figgie International	Northrop
Garrett	Raytheon
Gencorp	Rockwell International
General Dynamics	Sanders Associates
General Electric	Singer Company
Goodyear Aerospace	Sundstrand
GM-Delco Electronics	Teledyne
GM-Hughes Aircraft	Texaco
GTE	Textron
Hewlett-Packard	TRW
ITT	United Technologies
Lear Siegler	Westinghouse Electric
Litton Industries	

An identical survey was also sent to a sample of several hundred nondefense companies, representing a cross-section of service, financial, automotive, chemical, and other industries. Nearly 100 of these surveys were returned. The survey results for these nondefense industries will be mentioned only when there are sizable differences in the responses from the aerospace companies.

Indices of Quality
The first question, "Does your company have a specific 'statement of quality' (a quality-related credo to which the company adheres)?" was affirmatively answered by 87% of respondents. This was a typical response for all industries.

Figure 1 displays the results of the second question: "What are the indicators used to determine the status of quality in your company?" Respondents were asked to circle all of the indicators that applied to their companies, consequently the results total more than 100%. An average of five of the nine (plus "other") categories were used (compared with slightly more than three in nondefense industries). "Customer complaints" was the most frequently cited indicator of quality status in both defense and nondefense industries. More frequently cited in defense industries, however, were quality reports, cost of quality, external audits, and warranty costs. Given the introductory comments that quality reports and external audits

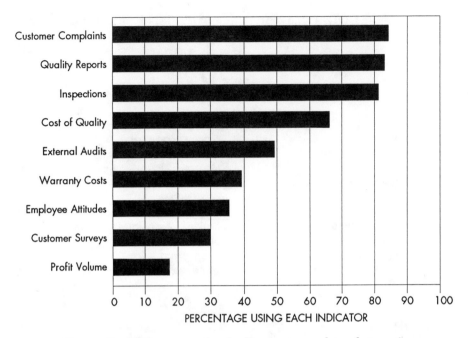

Figure 1. "What are the indicators used to determine the status of quality in your company?"

were more frequently cited is not surprising, although the latter was expected to be higher (49%).

Respondents were then asked how their companies define quality improvement. The results are displayed in Figure 2. Two of three respondents indicated that quality improvement is "product-based," whereas two in five indicated that it is "cost-based" or "process-based." Defense companies were more oriented toward cost-based and process-based definitions of quality improvement. Nondefense companies were more service-based in these definitions, as expected, because many of these companies are service-oriented.

Next, survey recipients were asked to rank the list of items shown in Figure 3 in terms of their company's priorities as ways to improve quality. For first place ratings, "change in company culture" was most frequently chosen, followed by "process control" and "employee education/training." Outside of the defense industry, the top three priorities were in reverse order: employee education, process control, and change in culture.

These executives ranked cost, schedule, quality, and profit by level of priority in their company, based on "senior management's actions, decisions, and emphasis." The results, indicated in Figure 4, show quality received the highest priority, followed by profit. Of lesser priority are schedule and cost. Later, the consistency of these evaluations will be examined.

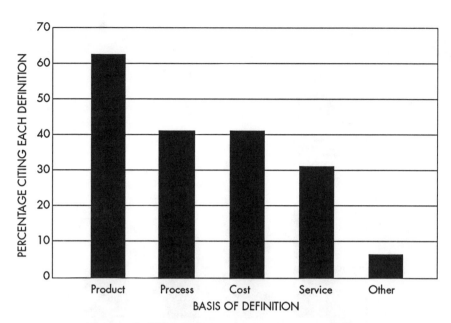

Figure 2. "How does your company define quality improvement?"

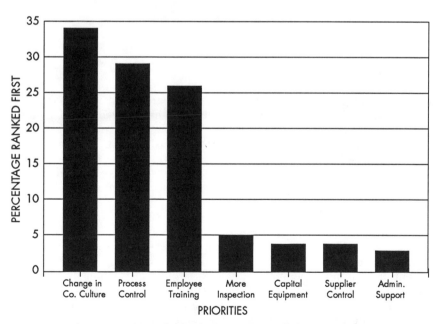

Figure 3. "Rank these in terms of their priority as ways to improve quality."

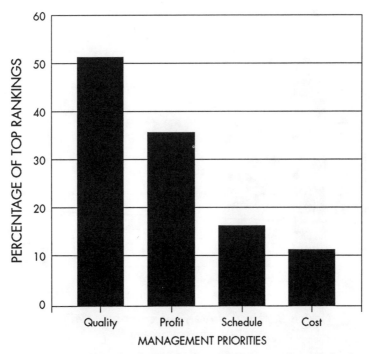

Figure 4. "Rank these four areas in order of their level of importance in your company"

Prime Causes of Quality Problems

"What do you feel are the prime causes for quality problems in your company?" Respondents were allowed to select any number from the list of items shown in Table 2. "Lack of adequate training" was most frequently chosen, even though on the previous question, this concept was ranked third after "change in culture" and "process control." "Management emphasis on schedule" was a close second, tied with "failure to follow established practices and procedures." "Customer schedule requirements" ranked fifth, lower than expected in terms of absolute numbers, but it was higher in defense than elsewhere.

Several causes of quality problems were ranked higher outside the defense industry—lack of personal accountability, lack of clearly defined criteria for quality levels, and conflicting standards of quality. Given the military's role as outlined previously, it is not surprising that these issues present more of a problem outside of defense. On the other hand, several items were more frequently cited within defense: customer schedule requirements, unrealistic contract specifications, and late delivery of vendor materials. The first two findings are consistent with this interpretation of the defense industry's uniqueness, as is the third item, given the financial penalties imposed for late deliveries of product.

Table 2
Prime Causes of Quality Problems

Percentage Citing Each Factor
51 Lack of adequate training
48 Management emphasis on schedule
48 Failure to follow established practices and procedures
38 Failure to communicate changes in design, specifications, etc.
34 Customer schedule requirements
34 Poor quality of vendor materials
32 Poor communication between levels of management
32 Lack of personal accountability
31 Late delivery of vendor materials
30 Company practices inconsistent with quality goals
22 Poor performers
22 Lack of clearly defined criteria for quality levels
22 Conflicting standards of quality
21 Lack of measures for quality
20 Inconsistent or unfair application of established practices
19 Unrealistic contract specifications
16 Vague/nonstandardized wording on reports/instructions, etc.
16 Low visibility of problems
16 Best equipment not available
14 Lack of manpower
10 Concerns about excessive cost
6 Inefficient facilities usage

Changes in the Past Two Years

Survey recipients were asked to indicate changes made in the past two years to support quality improvement in their companies. These results are indicated in Table 3.

These results are very positive and promising signs in the defense industry. The only "low" item was "allowing more time in plans so the job can be done properly." Apparently the defense industry does not regard this as necessary or relevant, but this will be addressed later.

These changes were not much different in nondefense industries in terms of rank-orders, but in terms of absolute percentage, defense contractors are instituting the following changes:

1. taking more specific actions to reduce the cost of quality;
2. using additional sources of data to better understand job processes;
3. increasing automation of processes; and
4. adding controls for health and safety, all on a more widespread basis.

Table 3
Changes in the Past Two Years To Support Quality Improvement

Percentage Citing Each Factor
- 84 More emphasis on prevention rather than correction
- 81 More on-the-job training
- 80 More emphasis on the critical few (versus trivial many) problems
- 79 More orientation of all new employees for quality improvement
- 78 New tools, instruments, or equipment
- 73 Specific actions taken to reduce the cost of quality
- 71 Additional training/controls for health & safety of employees
- 69 More formal training in quality improvement methods
- 68 Less scrap and "return for rework or repair," without reducing quality standards
- 62 Work-related documentation more readily available
- 57 Increased automation of processes
- 56 Additional data sources to better understand processes
- 55 Work processes redesigned/streamlined to reduce complexity
- 54 Increased attention to how each person's job fits into the whole to improve team action
- 49 Reorganization of job responsibilities to increase accountability
- 23 More time for planning so the job can be done properly

Roles and responsibilities. A series of agreement-format statements were presented to the survey recipients. The results are presented in Table 4.

Company executives perceived management as caring, communicating, attending to quality problems, and providing the means to achieve quality goals. There were no differences by industry. The role of employees in quality improvement seems less pronounced. Documenting and making available plans for maintaining or improving quality are not yet at what we would regard as "healthy" levels.

Quality problems appear to be handled in responsive and appropriate ways. Personal experience dictates that corrective actions are not always followed up to determine if they are effective, but the extent of this follow-up is greater in aerospace, at least in part because such action is mandated by the customer.

Many companies seem to use the special-project teams approach to quality problems. This approach is more effective than other quality-improvement methods. For special-project teams, management gives priority to quality problems and, an interdisciplinary task team is created to solve those problems. The team is provided the necessary resources such as training, equipment, space, time, and information. One team may define the scope of the problem and identify its causes, and another team may be tasked to solve the problem. The teams are given specific time limits to complete the task, management attends to their recommendations, and

Table 4
Role of Management in Total Quality

Percentage Agreeing with Each Statement
91 Management demonstrates an active concern for quality improvement.
82 Management discusses quality-related operations with employees.
76 Management provides the necessary means for achieving our company's quality goals.
65 Quality problems receive the highest priority from senior management.

ROLE OF EMPLOYEES IN TOTAL QUALITY

64 Employees are informed of the uses, performance, and problems of our products/services.
57 Control of quality is delegated to the operating level.
51 Each employee is held accountable for his/her own work.
43 Plans for maintaining or improving quality are documented and made available to all employees.

WAY IN WHICH QUALITY PROBLEMS ARE HANDLED

80 Remedial/corrective actions are followed up to determine if they are effective.
79 When addressing quality problems, the causes rather than the symptoms are focused on.
79 Serious quality problems are resolved by special project teams.
57 We have "alarm systems" (e.g., control limits) to inform us of process problems.

RELATED ISSUES IN TOTAL QUALITY

83 Periodic internal audits are conducted on the quality of products/services.
58 Poor performance is not tolerated.
51 Job process receives as much attention as the final product or service.
17 Our vendors are required to have training in quality improvement methods.

actions are implemented. Such approaches are more common in defense, as indicated by the survey results.

Internal audits are more frequent in the aerospace industry, as one would expect. Some concern may be raised about tolerating poor performance, attending to process as well as product, and controlling vendors. Dealing with poor performers—that is, counseling, documenting, and discharging—is an unpleasant task for the supervisor and is often avoided for that reason. Yet all companies must realize that there is a small percentage of personnel who are not performing up to expectations. Not only does this compromise the quality of products and services, it can have a serious impact on morale—to tolerate the poor performer may be worse than failing to acknowledge outstanding performance.

Until convinced that a quality product is the result of a quality process, the process will not be given much attention. Although the survey results indicate that contractors are "placing more emphasis on prevention than on correction," as indicated previously, the relatively low response to the item in Table 4, "Job process receives as much attention as the final product or service," suggests that the message "process determines product" hasn't yet been clearly understood. Finally, there seems to be little vendor control regarding quality improvement methods, even though "poor quality of vendor materials" was tied for fifth place as the prime cause of quality problems.

Relationships With Quality

An important aspect of quality improvement philosophy concerns the beliefs about the relationships between quality and cost, productivity, profit, and schedule. Recipients were asked whether they agreed that "improvements in quality result in increases in: 1) productivity, 2) cost, 3) profit, and 4) schedule." The percentage of respondents who agreed and disagreed are noted in Figure 5.

Almost everyone agreed that there is a positive relationship between quality and productivity—this was even more strongly agreed on in the aerospace industry than elsewhere. Many respondents also agreed that an increase in quality results in an increase in profit.

Fewer, but still a majority, agreed that quality improvements lengthen schedule requirements. It is possible that respondents misinterpreted this item, we intended to imply that efforts to improve product quality take time and that this time must be considered—thus, a planned increase in schedule projections. Perhaps this was misinterpreted, because an earlier

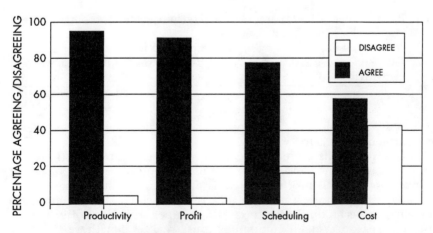

Figure 5. "To What extent do you agree that quality improvements result in increases in...?"

item already discussed, "allowing more time in plans so that the job can be done properly," received very low ratings, whereas most other topics did not.

Finally, only a slight majority felt that improvements in quality resulted in increased costs, whereas a sizable proportion of respondents disagreed. Increases in quality do result in decreases in material scrap and labor time associated with rework, consequently resulting in a decrease in costs. On the other hand, quality improvement is an investment in the future—it takes time and money, and more importantly, planning and commitment. Quality improvement is an indispensable investment which will result in increased profit. If by virtue of nothing else, quality improvement keeps companies in business.

Priorities Needing More Attention

In addition to the cost-schedule-quality trade-off, there are many other important aspects of total quality which must be considered. A number of these aspects are listed (Table 5) and questionnaire recipients were asked to indicate the top five they believed needed more attention in their company.

"Designing for producibility," number one on this list for defense respondents, was far higher than in other industries (where the number one item was "satisfying your immediate customer"). Other items that were ranked higher outside of aerospace were developing teamwork, reducing costs, and taking pride in your work.

**Table 5
Aspects of Total Quality Needing More Attention**

Percentage Ranking Item in Top Five	
43	Designing for producibility
42	Finding ways to improve work
39	Providing error-free work
38	Conforming to requirements
38	Satisfying the end-user
38	Developing teamwork
35	Reducing costs
34	Satisfying your immediate customer
34	Increasing productivity
33	Meeting schedules
32	Taking pride in work
25	Using resources efficiently
17	Using time efficiently
16	Reflecting good workmanship
12	Predicting time to complete a task
8	Enhancing reliability
3	Defining project responsibilities

Process Control Methods

Twenty-two different methods of process control were listed and are reproduced in Table 6. The percentages of companies using each of the methods are listed.

Table 6
Process Control Methods Used

Percentage Using Each Method	
85	Quality practices and procedures
84	Quality audits
82	Vendor ratings
79	Failure reporting
78	Design reviews
75	Problem identification/solving
71	Random sampling
71	Equipment maintenance
68	Quality costs
66	Statistical process control
65	Status reviews
58	Reliability and maintainability
55	Small group activities (e.g., QCs)
51	Systems safety analysis
44	Producibility
43	Work flow analysis
43	Experimental design
40	In-line environmental testing
35	Value engineering
30	Lifecycle costing
30	Finished product tear down
29	Quantitative decision making

All of these methods are more frequently used in aerospace, most of them twice as often. In particular, quality practices and procedures, vendor ratings, quality costs, and status reviews were used considerably more often. The major difference between aerospace and other industries in terms of the rank order of methods was that "quality practices" was first and "problem solving" sixth in aerospace, whereas the reverse was true in other industries.

Statistical Process Control

Two in three aerospace firms compared with one in two companies outside defense offer formal training to their employees in quality improvement. Nine in 10 aerospace companies (compared with five in 10 commercial companies) were familiar with statistical process control (SPC) methods at the time they received the questionnaire. Two in three aerospace companies have either offered courses in SPC or sent persons to be trained in SPC (compared with one in four commercially). Almost twice as many aerospace companies use SPC (87%) as nonaerospace companies (46%). Fifty-three percent of the aerospace contractors use SPC now and plan to expand its use in the future (17% in commercial sector), and only 13% do not use SPC nor plan to use it at this time.

SPC Training

In aerospace, more supervisors, managers, and technical professionals receive SPC training than those employees in production (Table 7), whereas in other industries, more production personnel receive SPC training than any other job category. The primary media of training are live instruction and textbooks, with more than half of the companies also using video instruction. The results for different media, shown in Figure 6, do not differ from other industries.

Table 7
Personnel Trained in SPC

Percentage of Companies Training	
67	Supervisors
64	Technical professionals
62	Managers
55	Production
38	Technicians
14	Business professionals
13	Maintenance

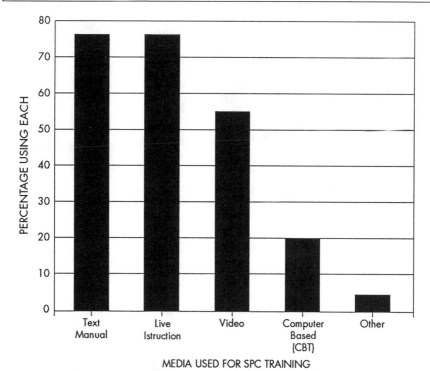

Figure 6. "Which media are used for training?"

Table 8 indicates the percentage of companies including each SPC topic in their training. Variable control charts, Pareto analysis, and histograms are the most commonly covered. Personal computer applications, scatter diagrams, and probability distributions are taught the least often.

Table 8
Topics Covered in SPC Training

Percentage Covering Each Topic	
78	Pareto analysis
78	Control charts for variables
71	Histograms
69	Sampling statistics
67	Control charts for attributes
65	Data stratification
58	Cause-and-effect analysis
58	Acceptance sampling
56	Guidelines for implementing SPC methods
55	Probability distributions
45	Scatter diagrams
33	Personal computer use in SPC

"Guidelines for implementing SPC methods" is comparatively low, given the importance of this topic. Several topics are covered more frequently in aerospace than in other industries, most notably data stratification, Pareto analysis, scatter diagrams, and histograms.

Half of those in hourly positions and half in salaried positions receive training at the companies offering SPC. Salaried personnel receive more training (36% with more than 10 hours) than hourly personnel (only 15% receive more than 10 hours).

SPC Administration

Three in five defense contractors are not required to use SPC which is similar to nondefense industries. Less than one in five aerospace companies require their suppliers to submit records of SPC use. One in three companies has an SPC steering or executive committee.

Table 9
Personnel Administering SPC

Percentage of Companies	
74	Quality
31	Production
15	Administration
11	Other
7	Training
2	Finance

Table 9 shows that quality personnel administer SPC most often (three of four versus two of four outside the defense sector). Quality Control personnel are also most likely to collect SPC data, more so than product operations personnel (Table 10).

Table 10
Personnel Collecting SPC Data

Percentage of Companies	
59	Quality Control
48	Operations
19	Engineering
11	Inspectors
11	Other

Other SPC Methods

Only one in four companies indicated that they use experimental design—for example, Taguchi methods—to pinpoint sources of variation. Most companies use variable control charts, even more in defense than nondefense (Figure 7). Nine of 10 companies use acceptance sampling (versus six in 10 outside aerospace), and of those who do use acceptance sampling, most use single acceptance sampling (twice as often in aerospace). Figure 8 shows the types of acceptance sampling used. Three in four defense contractors use MIL STD 105D for sampling (less than half outside defense use the military standard). Very few (12%) contractors identify critical dimensions that mandate SPC procedures of engineering drawings.

Highlights and Significant Results

There were five issues raised by the survey that require further consideration. They were perceived need for change in company culture, schedule (and cost) as driving factors, need for training employees, vendor control, and the role of employees in quality improvement. There were other issues or activities that stood out as strengths or presented further opportunities for improvement, but these five issues are too easily overlooked.

Change in Culture

The quality vice-presidents and directors in aerospace indicated that a "change in company culture" was the top priority to improve quality in their companies. This certainly raises several questions, in particular what characterizes their company cultures now, why do they perceive a relationship between their current culture and quality, in what direction do they want their culture to go, and how will they manage to change their culture? Given the nature of the survey, these questions can't be answered on behalf of the participating companies, but some clues can be offered.

The cultures of defense contractors have been shaped over the years, in large part by the customer, much like a child's personality is shaped by his or her parents. Sometimes contractors will reflect the customer's position—for example, if the customer behaves in such a way that

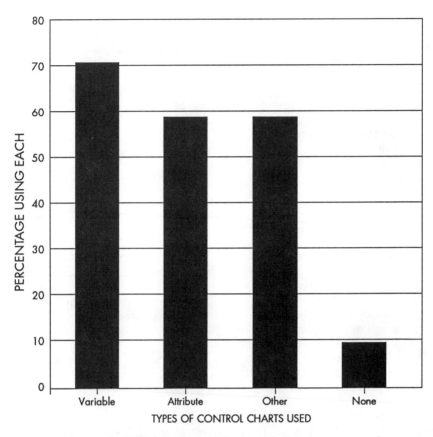

Figure 7. "Which types of control charts are used in your company?"

emphasizes schedule over all else, then contractors will place emphasis on schedule. If the customer pushes for the lowest possible cost and contractors must make a profit to stay in business, then an adversarial relationship begins.

Driving Factors
Are "schedule" and "cost" driving factors? When asked to rank order the importance of these factors, along with profit and quality, "based on senior management's actions, decisions, and emphasis," schedule and cost were clearly at the bottom of the list. But does this represent "espoused philosophy" or actual behavior? Clearly, "management emphasis on schedule" and "customer schedule requirements" were seen among the prime causes of quality problems, although "concerns about excessive costs" were not regarded as prime causes by these executives.

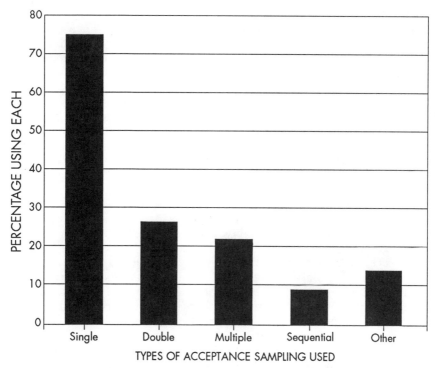

Figure 8. "Identify the types of acceptance sampling used in your company."

The only activity presented in the survey as a change to support quality improvement that was not endorsed by a majority of the survey respondents was "allowed more time in plans so the job can be done properly." The time needed to do a job properly and the time permitted to "just get it done" may differ, and plans should account for this. Few people regarded that "predicting time to complete a task" is a top priority needing more attention. They did, however, consider "meeting schedules" and "reducing costs" as more important.

Training Employees
"Lack of adequate training" was regarded as the prime cause of quality problems. If it is the primary cause of problems, then are employees getting more training, are they getting the appropriate training, is it quality training, is it just in time (JIT) training (that is, training provided just before they are asked to implement their new skills on the job), and are they using it on the job? Considering the responses to several questions, management is well on the road to eliminating this source of problems. Most respondents claim that there has been more orientation in quality, more on-the-job training, and more formal training in quality improvement methods during the past two years. The appropriateness and the quality of that training is unknown.

Vendor Control

"More control over suppliers" was not regarded as a high priority to improve quality, yet "poor quality of vendor materials" tied for fifth position as a prime cause of quality problems. "Vendor ratings" are frequently used (four out of five) as a process control method. Yet only one in six companies agreed that "our vendors are required to have training in quality improvement methods," and only one in six of the companies using SPC require their suppliers to use SPC. Almost all of these companies would agree that they are cited for poor control over vendor quality when they are audited by their military contracts administration group. Keeping good vendors and controlling their quality is one of the challenges that faces this industry.

Role of Employees

What is the role of floor personnel in quality? Several questions addressed this role in one way or another, and a few issues were problematic. "Employee attitudes" toward quality do not indicate to company executives the status of their own quality. "Poor performers" were not regarded by these executives as a prime cause of quality problems.

"Lack of personal accountability" is not considered as often a cause of quality problems in aerospace as in other industries, and "reorganizing job responsibilities to increase accountability" was the second least change occurring in the past two years to support quality improvement. Only half of these aerospace companies agreed that "each employee is held accountable for his or her own work," and slightly more than half agreed that "poor performance is not tolerated." Finally, almost as many disagreed as agreed that "control of quality is delegated to the operating level."

Quality gurus such as W. Edwards Deming and Joseph Juran note that "80% to 85% of quality problems are the responsibility of management," thus, they contend that employees play a minor role in quality improvement. More clearly stated, management is responsible for systems problems. What appears to be a laissez-faire attitude toward individual responsibility and accountability, however, is of concern. Workers are given little control over quality and are regarded as a minor component of the quality improvement effort. This appears to contradict earlier evidence citing lack of training as the prime cause, implying that employees are the cause of quality problems, albeit through no fault of their own.

FUTURE TRENDS

Some items on the survey do indicate some direction for "where we're going from here." For instance, most aerospace companies use SPC now and more than half intend to expand its use. These methods seem to be here to stay and are becoming an integral part of daily operations and management decision making, as they have been in Japan since the 1950s. Considerable emphasis will be placed on "designing for producibility," training our design engineers in these methods, increasing their familiarity with production processes, and increasing the engineering-manufacturing interface, which has long been a source of interdisciplinary conflict.

Other changes undoubtedly will occur as a result of the changing conditions under which defense contractors do business with the military. The critical feature of these changes, such as "cost-sharing," "streamlining," and "total ownership costs," is the impact on quality. Some of these changes focus more on improvements in quality—others offer further contraints on quality. Any changes that drive up costs without improvements in quality result in "less defense for the dollar"—a situation which neither the contractors nor the American public desire or deserve.

Chapter Four

Ford Quality Is Job Number One

Dean E. Smith, Jr.
Product Assurance
Ford Motor Company

Ford's recent successes in quality, sales, and profits have gained much attention in the past few months. Almost always when that subject is being discussed, the conversation quickly turns to the question, "What are the main reasons for the successful performance? What are the key actions that were taken?" And when the response is "Ford did a better job of listening to customers and responding creatively with products and service," quite often people are puzzled, as if there must be more than that simple explanation. But the truth is that the basic approach is simple—customers must be listened to and responded to in order to accomplish the following:

1. improve product quality;
2. improve product value; and
3. provide products that appeal to customers.

Although the approach is simple, the execution was more complex. It involved changing many business processes, modifying peoples' attitudes, creating a more team-oriented environment, and learning to become more comfortable with "change" instead of the usual ways of doing business. The remaining text is devoted to some of the highlights of this execution process.

Any change as dramatic as Ford experienced must begin at the top with a total commitment that is communicated throughout the organization and reinforced by upper management in their daily activities. Ford's commitment is best summarized in the "Company Mission and Guiding Principles."

Company Mission and Guiding Principles
Mission. Ford Motor Company is a worldwide leader in automotive and automotive-related products and services, as well as in newer industries such as aerospace, communications, and financial services. Ford's mission is to improve products and services continually to meet customers' needs,

allowing the business to prosper and to provide a reasonable return for stockholders, the owners of Ford.

Values. How a company accomplishes a mission is as important as the mission itself. Fundamental to success for a company are these basic values:

People: People are the source of strength. They provide corporate intelligence and determine reputation and vitality. Involvement and teamwork are core human values.

Products: Products are the end result of efforts and should be the best in serving customers worldwide. As products are viewed, so is a company viewed.

Profits: Profits are the ultimate measure of how efficiently customers are provided with the best products for their needs. Profits are required to survive and grow.

Guiding Principles

Quality comes first. To achieve customer satisfaction, the quality of products and services must be the number one priority.

Customers are the focus of everything done. Work must be done with customers in mind, and better products and services must be provided.

Continuous improvement is essential to success. Companies strive for excellence in everything: in products, in safety and value, in services, human relations, competitiveness, and profitability.

Employee involvement is a way of life. Ford is a team—the team must treat each member with trust and respect.

Dealers and suppliers are partners. The company must maintain mutually beneficial relationships with dealers, suppliers, and other business associates.

Integrity is never compromised. The conduct of a company worldwide must be socially responsible and command respect for its integrity and for its positive contributions to society. Doors must be open to men and women without discrimination and without regard to ethnic origin or personal beliefs.

This commitment has been communicated to all Ford employees and has also been shared with suppliers and dealers. There should be no uncertainty about Ford's purpose, values, or principles.

An important step in the day-to-day reinforcement of the company mission and guiding principles involves the "role modeling" by members of upper management. This is most evident through providing leadership which demonstrates that quality is the superordinate goal. Specific examples include the following:

1. devoting significant personal time to quality efforts;
2. making quality the top discussion priority; and
3. including "quality contributions" as a key factor in individual career progress.

In the environment created by a clear statement of company purpose and by day-to-day demonstrated leadership, the initial quality improvement efforts were directed at eliminating mistakes (rejects, warranty repairs, etc.) and improving conformance to specifications. This is a very natural and usual beginning to a quality improvement effort because it allows the organization to use the procedures and tools that are already familiar, tightens up the discipline, and usually provides near-term improved results, which maintain the enthusiasm and incentive and reinforce the process.

Although these efforts were under way, Ford recognized that to maintain the momentum and achieve "continuous improvement" would require a thorough review of the major business processes and the way employees interacted with each other as a team. This approach to "process improvement" is a continuing requirement as the basis for achieving continuous improvement. The changes that took place in the following selected examples are typical of the "process improvement" effort that has become a key ingredient in Ford's quality strategy.

Quality Objectives

Historically, quality objectives were expressed in end-of-line rejects, percentage of defective material, or warranty repairs. All of these measures were directed at "mistakes" and many at Ford's definition of "good or bad," not necessarily the customers. The measures also depended on measuring the outcome of an event "after the fact," and then responding with modifications or corrective action.

As a result of reviewing this "objective process," Ford now expresses its ultimate quality objectives in terms of "customer satisfaction." Thus the customer decides what's important (his or her expectations) and, based on the customer's perception of product performance, tells Ford how the company did. Focusing on satisfaction also helps Ford to not only reduce the aspects that customers don't like, but also to expand on the aspects they do like (bolster product strength) and plan added emphasis on dealership service, a vital factor in overall satisfaction.

The key is not that Ford restructured the quality objectives, but that the importance of "listening to the customer" was renewed and of building on strengths while improving in areas judged by customers as requiring attention. As an example, the shift to customer satisfaction as the key measure of quality caused people throughout the organization to become more familiar with Ford's marketing research information. More than 55,000 employees in North America were exposed by seminar and videotape to sources of customer information, how the information was gathered, and what it means. Ford also created a network of "customer information coordinators" in each organization to help users in their knowledge of the information and, through feedback, to make sure Ford was gathering "customer" answers to the right questions.

Product Development Process

An ongoing review of the product development process was initiated with an objective to significantly improve process and product quality and reduce program execution costs and time. The initial effort was directed at understanding the shortcomings in the current process and identifying the areas where improvements would have the most significant impact on the final results. Product development processes of other selected major companies were also reviewed to add additional insight into improved methods. Prerequisites to an effective product development process include the following:

1. a clear, concise, and ongoing flow of the "voice of the customer";
2. an overall business plan with complete up-front planning and analytical assessments with proper allocation of resources;
3. a product cycle plan that is customer-driven, technologically innovative, and manageable with available resources; and
4. ongoing development of technology, which is demonstrated to be capable.

With these important supporting elements, the product development process was defined to include the following key steps:

1. program initiation - based on approved cycle plans;
2. define product alternatives - both vehicle and system, also establish criteria for evaluating alternatives;
3. evaluate alternatives - vehicle and system versus selection criteria;
4. select best alternative - define assumptions and targets for selected alternative;
5. program approved - program approved, targets become objectives;
6. program progress reviewed - program assessed versus objective, verification prototype design frozen; and
7. production launch begins - engineering sign-off completed, functional build started.

Core teams of knowledgeable people were formed in each of these key areas to define and implement improved process steps required to support the overall objectives of better quality and less cost and time. An important factor in this effort was not only the revisions that took place in executing each process step, but also in defining measurable parameters to evaluate the "quality of the event." As the improvements are defined, they are "time-phased" into future vehicle and system programs currently approved and under development. The overall coordination and implementation of this process improvement effort directed at the product development process is being managed by a dedicated team of people responsible for implementing the process and assisting with interface issues between separate organizations.

"Team Taurus"
Examples of a team-oriented product development process are the very successful Taurus and Sable vehicles. Late 1979 and early 1980, when the program began, was a very difficult period because of a major recession

and resulting slow car sales. The company was losing millions of dollars and needed a major quality improvement. Ford needed a new way of doing business—of designing and developing products. The Taurus and Sable program that was evolving out of the conceptual stage would be a major beginning. They would be significant new products in a market segment that represents about 25% of industry volume. The focus had to be on the customer—Ford had to develop a better understanding of what customers wanted and how to maximize customer value. A major team effort was required. Earlier involvement of downstream activities was required. Finally, the program had to be flexible and move with the market to stay in step with a market that was changing rapidly. In summary, these were the factors that led Ford to develop a best-in-class philosophy and approach.

The results of the Taurus and Sable are volume and profits that are exceeding objectives. The Taurus is the number-two selling car in the United States behind the Escort. The Taurus and Sable are recognized as major new products and merited Motor Trends' "Car of the Year" award. Following are some of the specifics that were responsible for the successful outcome:

1. Ford recognized early in the program the need to develop a comprehensive listing of everything the customer looks at, feels, hears, and functions. In other words, the hundreds of details that make up the total perceived quality of a car. Ford's best-in-class-checklist included more than 400 design elements that went beyond reliability, durability, and fit and finish and focused on convenience and function.
2. Ford divided this list of 400 details into three categories:

    - ride, handling, steering, smoothness, performance feel, and comfort;
    - mechanisms and operating-type items that are effort-related; and
    - overall system, complete vehicle, and man/machine interactions.

3. Ford selected and obtained some of the best-in-class cars in the world—not just in the United States—and had engineers, stylists, designers, and product planners review and evaluate these cars in all of the categories described previously. This evaluation allowed Ford to determine, of all of the 400 detailed items, what represented the best-in-class approach currently available.
4. The company reviewed the status of Taurus and Sable designs with an objective of being the best. In appropriate areas, the design or process was changed. However, it is not possible to be the "best" in everything. Sometimes trade-offs are required between competing items, but in every case Ford reviewed the complete details and made the ultimate decisions with the customer in mind. For most items, the end results were best-in-class, which, from a competitive point of view, is outstanding considering that Ford identified 29 different cars for the various best-in-class details.
5. To verify the success of this approach, Ford conducted market research on early Job #1 prototype 10 months before introduction. Potential

customers were asked to evaluate the Taurus and Sable along with competitive cars. The results were used to determine the customer reaction to products and to identify any particular shortfalls that could still be improved. The research results identified a winner, and subsequent evaluations confirmed improvement in areas where changes were made.

6. Another very important aspect of the approach was total participation by all involved: suppliers, dealers, service, sales, marketing, engineering, manufacturing, and assembly. Ford conducted design reviews on every system and component in the car. The company went completely through the car, part by part, three different times, during a period of three years. These reviews were led by the vice president of product development and the Team Taurus leader. This team was a living example of upper management participation and reinforced quality and attention to customers. During these reviews, Ford never made a decision based solely on cost—decisions were always based on the expected customer benefit and the functional plusses. Ford missed the cost objective in the program, but the results were worth it. They proved again that if you take care of the customer, the customer will take care of you, and the sales and profits will take care of themselves.

This approach and continuing refinements are being expanded to all future vehicle and component programs. Ford's objective is to become the best quality, lowest cost automotive producer in the world. These are just three examples of the actions being taken to help Ford reach the long-term objective.

MANUFACTURING FOR QUALITY

Let's review some of the changes taking place in manufacturing that reflect our customer-driven philosophy and which will help us reach our customer satisfaction objectives.

Manufacturing is a key arena in the drive for better quality for many reasons. Manufacturing incorporates the efforts of engineering, styling, development, procurement, and tools/facilities, and the finished product is born. And, as such, manufacturing is the arena where unforeseen interface problems materialize. Secondly, manufacturing is the final step in the process before the product is sent to the dealer and retail customers.

Ford's fundamental approach to improving manufacturing's contribution to quality is based on the following: 1) much earlier involvement in future programs to ensure manufacturing needs are satisfied and strengths are used; and 2) more intensive before-the-fact prove out of the manufacturing process, tooling, and facilities. The following is a list of examples of the actions and tools being used to improve manufacturing quality.

Management quality team. Teams were formed for each major component of the product within the overall body, chassis, electrical, and powertrain systems. The teams include engineering, manufacturing/assembly, supplier, and service representatives. The teams are responsible for understanding and resolving current quality concerns and feeding the knowledge gained upstream to prevent recurrence. This forum provides for interaction among various disciplines and an entry into the forward planning of future products. The individual teams report progress on a regular basis to the management quality team, which includes the management of the organizations involved.

Earlier manufacturing involvement. Examples of earlier manufacturing involvement were evident on the Taurus/Sable program and include the following:

1. Manufacturing and assembly people at all levels participated in engineering design and development reviews and evaluated early vehicle prototypes. A total of 550 ideas that came from this involvement were incorporated into the program to facilitate manufacturing quality as a contribution to final vehicle quality levels.
2. Production suppliers for major components were identified early in the program and participated in the design and processing development. Additionally, a series of visitations were conducted with major outside suppliers to coordinate the development process, reach timely decisions, and ensure that prototype parts were available which fully represented production levels.

Process capability studies. More extensive process capability studies are being conducted well in advance of production. These studies evaluate the tooling, machinery, and man/machine interface to help establish process stability and provide the knowledge required to support improvement in process capability.

Statistical process control. A comprehensive training program was developed and implemented to upgrade the qualifications of all levels of employees on statistical and logical thinking. Statistical process controls are also being implemented in critical processes throughout manufacturing and assembly.

Design of experiments. This approach is being used to gather quantified data on the variation and interaction of variables in selected process areas. Selection is based on customer sensitivity and the consequences of the interacting variables. For example, for doors, the interaction of variables must be managed and the design and process must be made more robust to achieve the proper balance between door efforts/feel, water leaks, and wind noise.

Customer liaison personnel. Technically qualified personnel were added to assist the quality management team structure in obtaining detailed descriptions of customer concerns by evaluating customer vehicles and observing dealership repairs in major geographical locations. This knowledge is used to correct current concerns and is fed back into future designs and processes.

These actions are a sampling of the kinds of approaches being used to facilitate "manufacturing for quality." A manufacturing employee might ask, "What approach can I use in my area of responsibility to improve my quality contribution?" Employees should consider the following process, plan their actions, and get started. The most difficult aspect of the whole process is taking the first step. The process includes the following steps:

1. making a commitment;
2. knowing the customer;
3. communicating with the customer;
4. understanding the customer requirements;
5. strengthening actions; and
6. measuring performance.

The process is simple—there is no magic. It involves paying attention to very basic elements. The question to ask is not do I understand these factors, but how well am I following them? Do I really know what my customer requirements are, am I fulfilling them, or am I providing products or services that meet my interpretation of my customer's requirements?

Any change must begin with a commitment to change. Change is not comfortable for most—most people tend to continue to operate as usual because it lets them remain in their comfort zone. If an employee is serious, he or she will make a commitment to improve performance by 50% or to let the internal customer write his or her next personal performance review. Without a strong commitment, the rest of what a person chooses to do will be largely "going through the motions," and each time a change is necessary, that person will rationalize to continue acting as always—because it's comfortable.

Do you know who your customers are? They are the people who buy the final product and certainly your immediate boss. But the people in the next department in the process, the people who depend on your output to do their job, are also customers. Concerning your customers, do you know what "drives" their satisfaction? It is your effort to provide products, services, and information beyond what the customer "expects" and the responsiveness you demonstrate when they come to you for help or information. Those are the elements that set the upper limit on your customer's satisfaction.

The mistakes you make, or in their terms, "the things you do wrong," detract from their satisfaction. And the extent to which you can eliminate those mistakes will minimize dissatisfaction. But if your main focus is on eliminating "things gone wrong," even if you're totally successful, you'll end up being a "middle-of-the-road" performer.

Everyone is a customer every day. Think about the last time you went to a restaurant or made a purchase that really impressed you as a satisfying experience. Was it because the waiter or salesperson didn't do anything wrong? No, they are expected not to do anything wrong. It was because they did something right, beyond your expectations, and they went out of their way to serve your needs.

Communicate with customers and understand their needs in terms which you can act on. If you provide the information to another internal department that they need to do their job, is it in a format which they can readily use? Or do they have to spend time analyzing your information and rearranging it? Is it timely? Does it support their time-phased decision point? Is your information complete or does it leave unanswered questions? What would you have to do differently to answer their questions?

Share your new ideas for different and better ways to do your job with your customers. Will it serve their needs better?

Strengthening action is fundamental to increasing the satisfaction of customers. Most people are faced with the competitive challenge to operate more efficiently and more effectively, and that translates into not only doing things right, but doing the right things. The "forgetting curve" is more important in this process than the "learning curve." Examine carefully the things in the past. Identify the elements that have outlived their usefulness or provide marginal value. Stop doing outdated tasks—it's the only way you can make room for new, more productive, and meaningful methods. Be willing to take a "risk" and move forward—"nobody ever stole second base with their foot on first."

The last step is to measure your performance. You and co-workers will know how well you're doing, but also "ask your customer—use one-on-one conversations and use customer surveys—with internal customers, as well as the purchasers of your final products and services. Use your findings to continually sharpen your focus and guide your future actions.

When considering the examples of quality improvement efforts reviewed, the question arises, "What is the quality improvement model that is being used at Ford?" Ford does not have any specific, step-by-step model that will produce the results Ford's striving for—there isn't any "magic button." The Ford quality improvement effort starts with a corporate commitment, involves all of Ford's people, and includes examining and improving everything done, every process in the business. The effort involves attention to detail and spans across product development, dealer service, accounting, personnel, training and education, food service, supply, and distribution and reaches from the board room to the shop floor. A "generic" model of this approach includes the following:

1. thoroughly understanding the current process;
2. examining the steps that need improvement;
3. developing and testing the improvements; and
4. incorporating the improvements.

The final step is to continue to improve every process, no matter how well it is currently performing. Make continuous improvement a way of life—the results of this approach are evident in terms of product quality, productivity, and profitability.

In summary, a lot of different changes are happening at Ford, and they all contribute in some way to improved overall quality and the successes in sales and profits. But what about the future? How does Ford plan to continue to build on the momentum? Note that in the beginning it was stated the basic approach is simple, even though the execution was more complex. At Ford, we plan to continue the approach that has proved to be successful:

1. improve product quality;
2. improve product value; and
3. provide products that appeal to customers.

Ford will continue efforts under way currently, emphasize "listening to the customer," and continue to examine and search for ways to improve all business processes.

Chapter Five

Managing for Quality Improvement In the Eastman Chemicals Division

Paul Hammes, John Wallace and Lee McConnell
Tennessee Eastman Company

INTRODUCTION

The Eastman Chemicals Division (ECD) of the Eastman Kodak Company intensified its quality management effort in 1983 by forming a group of quality professionals who were well-versed in quality disciplines. This group was instructed to develop a Quality Policy and a training program for people to implement this Policy.

In 1984, the company began training manufacturing personnel. By the end of 1985, approximately 90% of manufacturing's supervisors and operators had completed their quality training. Middle and upper management received three weeks of quality training, staff personnel received two weeks, shift supervisors received one week, and operating-level personnel received four days. This training included quality control, statistical process control, performance management, and teamwork concepts. Nonmanufacturing personnel began their training in mid 1985 and will continue into 1988. Realistically, quality training efforts will never end. Currently all employees are significantly contributing to the quality effort.

All company personnel now understand and are helping to implement Eastman's Quality Policy. This policy's primary purpose is to focus and direct each individual in making decisions in regard to producing a quality product (regardless of what that product is). The company's goal, as stated in the Quality Policy, is to be the world leader in product quality and value. This goal will be achieved by following two basic principles and eleven supporting elements.

The first principle is continual improvement. The present level of performance must always be improved upon. Most of these improvements will come about gradually as the people operating the process work together to find ways to make the process better. Initially, at least, continual improvement primarily involves identifying causes of process variation. This identification may appear somewhat simplistic and not particularly lucrative. However, at least 90% of Eastman Chemical Division's (ECD's) processes are capable of meeting customers' needs if they are controlled within their statistical capability.

The second principle of the Quality Policy is defect prevention. Producing defective products and sorting them out later is expensive. No inspection process is 100% effective at finding defects, and even good systems are only about 90% effective. Thus, anytime defective products are made some of them are shipped to customers.

The eleven supporting elements specify what must be done to achieve the quality goal. These elements serve as guidelines for managing quality within the company. Elements two through eleven are driven by element one, which states that quality improvement must be led by the line management to have any chances of success. The remaining elements deal with various aspects of continual quality improvement. Most of the elements deal specifically with leading employees in the integrated application of statistical process control, performance management, and teamwork (see Table 1 for Eastman Chemical Division's Quality Policy). Each of these will be explained briefly.

STATISTICAL PROCESS CONTROL

Statistical Process Control (SPC) is the application of statistical methods to attain and maintain control of a process. Whenever a specific product's quality is measured—either by testing or by some in-process measuring device—that measurement method should have a statistically valid way of detecting an assignable cause of variation. A control strategy or response sheet (checklist) is then used to systematically determine the cause of the abnormal variation. After it is identified, the cause can either be corrected or compensated for. One important note is that one must always attempt to correct and eliminate a source of abnormal variation and reserve compensation as a last resort.

Most of the control charts used within ECD, and in the chemical industry, are X charts. These charts are used because, generally, there is no reason to form subgroups—measurements are usually independent of one another. Subgrouping data slows down response time unless a moving average is employed. Process capability is estimated by using a two-point moving range, and outlying ranges are discarded. Most charts are modified Shewharts' with three sigma limits.

One unique aspect of SPC at ECD is the response sheet, which is basically a trouble-shooting checklist to help search for assignable causes. The sheet also allows the operator to document what is found during an investigation. Potential causes are listed in order of likeliness of occurrence. If an assignable cause isn't found, the response sheet contains guidelines

Table 1
Eastman Chemicals Division Quality Policy

Quality Goal
The leader in product quality and value.

Principles
1. Continual improvement — The current level of performance can be improved upon.
2. Defect Prevention — Prevention is more cost effective than detection and sorting.

Operational Policy
Establish process capabilities.
Control of every process to the desired target within the process capability. (This is to be given top priority.)

Elements Which Support the Principles and Enable Achievement of the Quality Goals
1. Management leadership and involvement.
2. Use of Statistical Process Control methods and techniques.
3. Manufacturing information systems.
4. Training of all personnel in job functions, quality responsibilities, quality disciplines, and statistical methods as appropriate for their job assignment.
5. Teamwork among all functional areas and organizational levels.
6. Long-term relationships with customers and suppliers.
7. Involvement of all employees in continual improvement through annual improvement goals.
8. Quality cost systems.
9. Short-term decisions consistent with long-term goals.
10. Performance standards that focus on both quality and productivity.
11. Competitive quality intelligence systems.

for taking compensating action. Response sheets are developed by teams, including operators, shift supervisors, area supervisors, technical people, and a facilitator—they are the key to a successful SPC effort.

PERFORMANCE MANAGEMENT

Performance Management (PM) is creating an environment that fosters the use of SPC and teamwork. PM involves creating positive consequences to reinforce desired behaviors. The PM system is designed by measuring and plotting a performance index, developing a plan of action to improve performance, setting goals, and having a plan for reinforcement when goals are achieved. PM provides a structure within which SPC and teamwork efforts can succeed. It forces management involvement, measurement of performance, and celebration of successes.

PM is based on the premise that people do what their management and peers reinforce them for doing. If performance measures focus on quality, people will focus their improvement efforts on quality. As people are reinforced for serving on teams and implementing SPC systems, they will do these tasks more often. As management and workers spend more time together at team meetings and celebrations, barriers such as fear and lack of communication will be broken. Management and workers develop measures and plans of action together and they celebrate their successes together. This allows efforts to be successful and makes them a part of the corporate culture.

Teamwork involves all employees in decision making, problem solving, and goal setting. Teamwork is used to develop, implement, and maintain SPC and PM systems. Teamwork, or employee involvement, is crucial to Quality Management because those who know the production systems well are the operators using the systems daily. Operating personnel too often have been an untapped resource, and teams provide a structure to use this resource to uncover and solve problems and make system changes. Through teams, people can now design and improve the systems they work within. Teamwork results in a commitment level far better than the mere compliance produced by past management styles.

There are basically two types of teams within the ECD. There are "natural unit teams," which consist of a supervisor and the supervisor's employees. These teams interlock from the top of the company to the bottom. Generally, they meet once per week and make decisions that deal with managing the business. Previously, these decisions were made by the supervisors, with little input from their employees.

A second type of team is a problem-solving team. This team is structured around a very specific, well-defined problem—team makeup is dictated by the problem. Usually a team includes shift supervisors and operators from 2 to 4 shift groups, an area supervisor, a staff technical person, and a facilitator. Meeting frequency and duration are also dictated by the problem. A unique feature of these teams is that more than one shift group is involved, thus shift-to-shift differences in experience and methods can be detected. This multigroup approach spurs all four shift groups to support and implement the team's recommendations. Most of ECD's SPC and PM systems are designed by this type of team.

QUALITY MANAGEMENT - A CASE STUDY
The following section illustrates how ECD applied the three facets of the Quality Management triangle (SPC, PM, and teamwork) on an actual production quality problem.

Introduction to Project

The Film and Fiber Esters Department, Cellulose Esters Division, Tennessee Eastman Company, produces cellulose acetate that is the raw material for filter tow. Filter tow is sold to cigarette manufacturers who use it to produce cigarette filters. One of the most important quality parameters of cellulose acetate is unreacted cellulose, which must be filtered out before filter tow is produced. Ideally, cellulose acetate should contain no fiber. The objective of this project was to minimize the level of unreacted cellulose fiber found in the cellulose acetate.

Cellulose acetate is produced by the solution method with an acetylation mixture of acetic anhydride, cellulose, sulfuric acid as a catalyst, and acetic acid as a solvent. The acetylation is heterogeneous at first, with the cellulose fibers suspended in the liquids. As the acetylation reaction proceeds, the fibers gradually lose their fibrous structure and finally pass into solution in the solvent to form a viscous syrup (dope). Because of the heterogeneous nature of esterification reaction, the cellulose fibers must be properly conditioned before esterification to ensure uniform diffusion rates of the chemicals into the fibers. The amount of unreacted cellulose fibers remaining after acetylation is a measure of the reaction efficiency. The dope is then hydrolyzed and precipitated into a solid form, which is washed and then dried. The dried cellulose acetate is then processed into filter tow.

This acetylation process has three in-process measures to evaluate the fiber level of the dope: a fibermeter, which gives a continuous readout; visual inspection made hourly by the operator; and fiber gradings of the collection tank sample of two systems that represent two different acetylation sources (see Figure 1). The grading of the fiber level is done visually by comparing a sample of dope to a lab standard. These standards are representative of fiber levels of 55, 110, 165, 220, 330, 440, and 550 ppm. This project concentrates specifically on attaining control of the collecting fiber samples.

Grading the collecting tank fiber levels in the two systems began in the spring of 1985 and was used primarily to investigate fiber level upsets in the final product. In October 1985, run charts with superimposed reject limits were installed. Whenever the fiber level of either system was above the reject limits, the dope was rejected, and the shift supervisors and area supervisor investigated the cause of the high fiber level. Each management level looked at the different potential sources of high fiber corresponding to their area of responsibility. This marked the beginning of an informal control system.

The Team

In March 1986, a team was formed to standardize the control system for collecting fiber level. The team included the area supervisor, two shift supervisors, two acetylation operators, an activation operator, a chemical

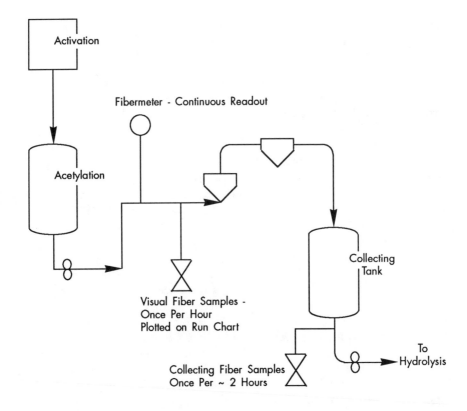

Figure 1. Fiber Control System

engineer, and an industrial engineer. This team continued to collect data that was to be used to develop statistical process control limits.

A typical run chart in March 1986 for System 2 shows that the fiber level is in good control at a very acceptable level (see Figure 2). The chart shows all the fiber levels at 55 ppm and 110 ppm, with the reject limit of 500 ppm. The run chart for System 1 shows that its fiber level is much higher and significantly more variable than that of System 2 (see Figure 3). More precisely, the average fiber level for System 1 was 200 ppm versus an average fiber level for System 2 of 60 ppm.

The next step was to develop a SPC chart for System 1 (see Figure 4). This control chart was the data's mean, upper control limit, and lower control limit on the chart. The periods of high fiber are much more frequent, signaling a definite difference in the two systems.

During its initial meeting, the team used the brainstorming technique to identify potential causes of high fiber. These causes were then integrated into a cause-and-effect diagram to categorize their effect on high fiber

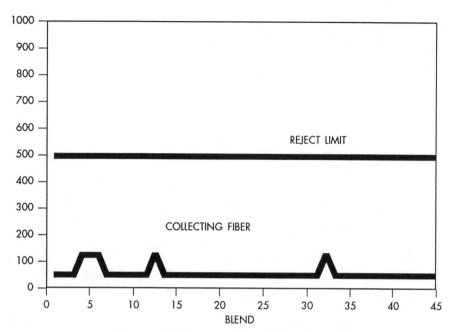

Figure 2. System 2 - Collecting Fiber

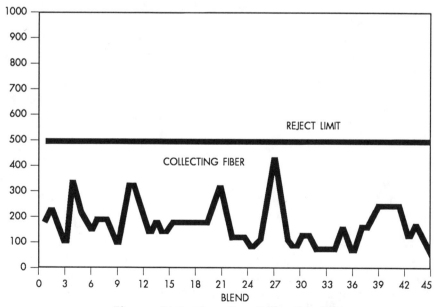

Figure 3. System 1 - Collecting Fiber

Industry Applications

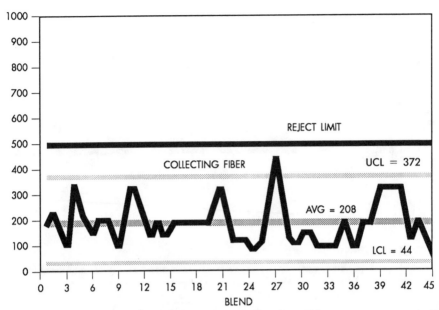

Figure 4. System 1 - Collecting Fiber

levels. The team identified 34 causes of high fiber, which were then given priority based on their category on the cause-and-effect diagram (see Table 2). The reason for ranking the causes was to give the potential causes priority according to their perceived probability of occurrence. An investigation action list was developed using the priority list of potential causes. This investigative procedure is called a response sheet.

These response sheets had some very unique features, one being that no compensating action could be taken for high fiber. Any compensating actions for high fiber would result in unacceptable viscosity degradation of the ester. The problem could only be solved by looking at all possible causes, finding the actual cause, and correcting it. To produce an ester with low fiber, production had to be done correctly the first time. Another unique feature of the response sheet is that a separate response sheet was developed for each level of responsibility in the process. The acetylation operator, the activation operator, the shift supervisor, and the area supervisor each have their own response sheets. The hierarchy of control places the responsibility for investigation and correction at the proper level.

After the response sheets were developed, it was time to identify what made System 1 produce dope at a higher fiber level than System 2. As stated earlier, System 1 had an average fiber level of 200 ppm, with only 6% of those values falling at the lowest standard level of 55 ppm (see Figure 5).

Table 2
Potential Causes of High Fiber

Priority*		Cause	Type					
Rank	Average Rating		Human	Machine	Materials	Methods	Environment	Other
1.	10.0	Reactor Temperatures Out of Control	X	X	X	X		
2.	8.9	Pretreat Spray No Uniform		X				
3.	8.8	Rate Changes and Start-Ups				X		
4.	8.7	Scale Flaps Leaking		X				
32.	2.9	Samples Mixed Up	X					
33.	2.5	Weather					X	
34.	2.3	Low % Ac_2O			X			

* Based on Significance and Frequency

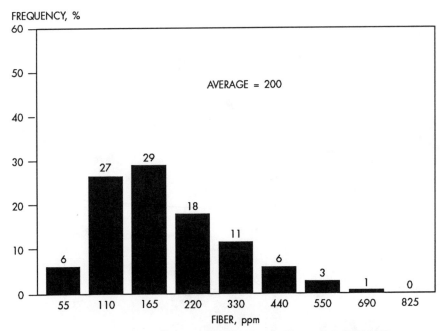

Figure 5. System 1 - Collecting Fiber - 3/17/86

The team—and especially one shift supervisor on the team—began a search. The shift supervisor noticed that one of the hammermills in System 1 did not produce as small a cellulose chip as did the hammermills in System 2. More teeth were then added to the hammermill so it would break the cellulose sheets into smaller fiber chips. This breakthrough improved the system's ability to acetylate the cellulose, thus allowing the average fiber level to decrease from 200 ppm to an average of 120 ppm. The percentage of 55 ppm gradings increased from 6% to 35% (see Figure 6).

The next system improvement was in the acetylation operation. The same shift supervisor noticed that System 2 had a faster reactor motor than System 1. The slower reactor in System 1 was replaced with a motor similar to the one in System 2. A significant improvement in the fiber level of System 1 was again realized. The fiber level decreased from the average level of 120 ppm to an average of 90 ppm, with the percentage of 55 ppm gradings increasing from 35% to 58% (see Figure 7). Interestingly, according to the motor specifications, both systems should have had the same motor speed as that of System 1. Thus, by mistake, a motor was installed that in actuality produced lower fiber levels.

Even though these breakthroughs were found by only one team member, the entire team was credited with this improvement because the

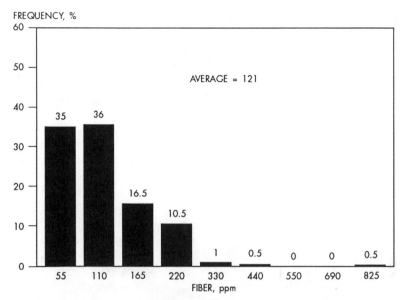

**Figure 6. System 1 - Collecting Fiber Smaller Cellulose Chip Size
4/1/86 to 4/17/86**

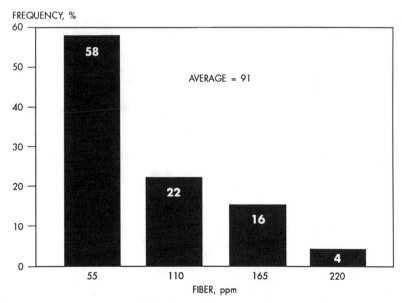

**Figure 7. System 1 - Collecting Fiber Increase Agitation in Reactor
4/17/86 to 5/6/86**

team discussions focused on the importance of agitation and prompted this individual to find the systems' differences.

After these improvements, the team decided to start designing their PM system to accompany their existing control system (control chart plus response sheets). This PM system would provide everyone with information on how they were doing, would help set improvement goals, and would specify what type of reinforcement everyone would receive when the goals were met. The team designed a scoreboard for performance feedback (see Figure 8). Each week the average fiber level for each system was plotted. These scoreboards were mounted in three places in the operating areas so everyone could readily see the system's performance during the past week. Next, the team established improvement goals for each system. These goals were carefully chosen to ensure they were neither too easy to attain nor so high they couldn't be reached. And finally, the team developed a reinforcement plan. This plan clearly stated what the reinforcement would be depending on which goal had been achieved. The plan also included who was going to receive the reinforcement and share in the celebration. The team wanted to include everyone who had a part in helping achieve these improvements.

The team's next task was to work out the details of transferring the control system technology to the other two crews. At ECD, the operating

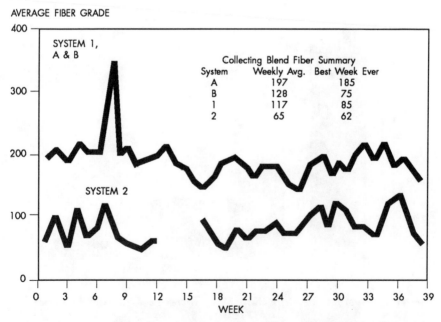

Figure 8. 1986 Product Fiber Feedback

level employees, including shift supervisors, work on a four-crew, rotating shift to run the processes 24 hours a day, seven days a week. Because two of the four crews were represented on this team, details of how to train the other two crews needed to be resolved. The team decided that the two supervisors on the team would train the other two supervisors, and each supervisor then trained his or her crew. This method of technology transfer worked extremely well because it enabled the other two supervisors to share ideas and make suggestions for further system improvements. This opportunity for employee input created a great deal of commitment to this new system. All four crews could now work together to reduce the fiber level in the ester. The control system has six elements: a control chart for both Systems 1 and 2, response sheets for the acetylation operators, activation operators, shift supervisors, and the area supervisor.

The control chart for System 1 collection tanks has, as its upper limit, a fiber level of 230 ppm with no lower limit required (see Figure 9). Fiber levels have been substantially reduced because employees are now investigating fiber levels greater than 230 ppm, which used to be the normal level of operation. Five points were out-of-control during the first few days after implementing the control system. By following the prescribed investigative procedure of the response sheets, assignable causes were found for three of the out-of-control situations.

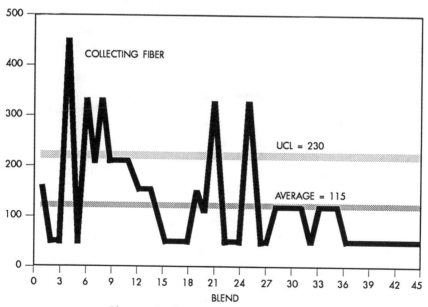

Figure 9. System 1 - Collecting Fiber

Industry Applications

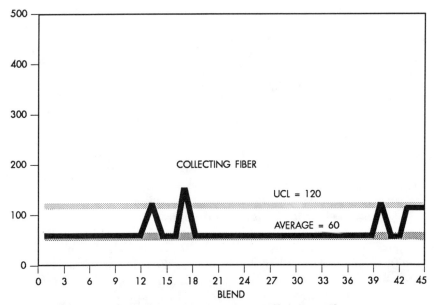

Figure 10. System 2 - Collecting Fiber

The control chart for System 2 collecting tanks has, as its upper control limit, a fiber level of 120 ppm with no lower control limit (see Figure 10). Immediately after the control system was implemented, the control chart signaled an out-of-control fiber level. The assignable cause was found for this abnormally high fiber level.

The acetylation operator's response sheet has three separate sets of responses, each corresponding with one of three different categories of the high fiber problem (see Table 3). The categories are a high fiber meter reading with normal acetylation temperatures, a high fiber meter reading with an acetylation temperature upset, and a normal fiber meter reading. Questions at the beginning of the response sheet, when answered, direct the investigator to the proper category of response. This procedure simplified the investigation because it quickly pinpoints the number of potential causes that need to be investigated.

The activation operator's response sheet is less complex than the acetylation operator's sheet (see Table 4). When fiber levels are out of control, the acetylation operator will automatically notify the activation operator, who can also check the activation process. The activation officer immediately documents the activator number, cellulose type, and lot number. He then checks the controls, flow controllers, and control charts for the activation operation. Finally, the operator checks the equipment to see that all important components are operating properly.

Table 3
Acetylation Operators High Fiber Response Sheet

Date _____ Time _____ System _____ Blend No._____
Fiber_____ Collecting Time_____ to_____
This is a response sheet to:
_____ One point above the upper limit
_____ Seven consecutive points above target

Investigate problem, identify cause, and take corrective action.
Circle Y for each item that was the cause.

1. Was the fiber meter reading HIGH when blend was collected?   Y/N
   If no, go to Step 9.
2. What was the visual reading this time? _____
3. Was the problem corrected?   Y/N
4. Were reactor temperatures normal?   Y/N
   If yes, continue. If no, go to Step 7.

HIGH FIBER METER/NORMAL REACTOR TEMPERATURES
5. A. Was acetylation catalyst out of control (low)?   Y/N
   B. Was there a pulp type change?   Y/N
   C. Have tower levels been low two lights?   Y/N
   D. Was there any dope rejected during collecting times?   Y/N
      Reject time(s)_____
      If so, did collecting level rise during reject time(s)?
   E. Was filter pressure abnormal?   Y/N
      If so, change filters.
   F. Check total acetic acid set points.
      Was it different than the parameter sheet?   Y/N
   G. Check ice take-offs. Were they choked?   Y/N
   H. Was there a viscosity/fiber relationship problem?   Y/N
6. Stop - Close Response Sheet.

HIGH FIBER METER/REACTOR TEMPERATURE UPSET
7. A. Was there a rate change or startup during collection time?   Y/N
   B. Was there any dope rejected during collecting times?   Y/N
      Reject time(s)_____
      If so, did collecting level rise during reject time(s)?   Y/N
   C. Check scale flaps for buildup and rams for adjustment. Problems?   Y/N
   D. Is there a viscosity/fiber relationship problem?   Y/N
   E. Was there a tower weight shift?   Y/N
   F. Was there under or over weight pulp?   Y/N
   G. Check ice take-offs. Were they choked?   Y/N
   H. Check conveyor feed roll speed. Problems?   Y/N
   I. Were reactor drives working incorrectly?   Y/N
   J. Was there any indication of water in the system?   Y/N
   K. Was there abnormal variability in anhydride temperature?   Y/N
   L. Was filter pressure abnormal?   Y/N
      If so, change filters.
   M. Check acetic acid and anhydride set points.
      Were they different than the parameter sheet?   Y/N
8. Stop - Close Response Sheet.

FIBER METER NORMAL
9. A. Did the visual fiber readings indicate the fiber meter was not working properly?   Y/N
   B. For C blends, was any dope collected from CA-2 or for D blends, was any dope
      collected from CA-1?   Y/N
      If so, was there a difference in fiber levels between CA-1 and CA-2?   Y/N
   C. Is there fiber in the acid finals?   Y/N
   D. Are the throttle bushings leaking excessively? (D system only).   Y/N
Response Sheet Closed - Initials_____

Table 4
Activation Operators
High Fiber Response Sheet

Date _____ Time _____ System _____ Blend No._____
Fiber_____ Collecting Time_____ to_____
This is a response to:
_____ One point above the upper limit
_____ Seven consecutive points above target

Activator No._____ Cellulose Type_____ Cellulose Lot No._____
Activator No._____ Cellulose Type_____ Cellulose Lot No._____

1. Do any of the following charts show anything abnormal?
   a. HOAc Flow   Y/N
   b. H_2SO_4 Flow   Y/N
   c. Pretreat Flow   Y/N
   d. Pretreat Tank Level   Y/N
   e. Activator Catalyst (out of control low)   Y/N
   f. Linear Cellulose Sheet Speed (out of control high)   Y/N

2. Do these conditions exist on the activators?
   a. Abnormal activator pressure   Y/N
   b. Clogged spray bars   Y/N
   c. Clogged wet plate   Y/N
   d. Leaking flexible hose   Y/N
   e. Buildup on activator   Y/N
   f. Bad bearings   Y/N

Response Sheet Closed - Initials_____

The shift supervisor's response sheet is completed only after the potential causes in the acetylation and activation operations have been thoroughly checked (see Table 5). The supervisor may change the activator on which the cellulose sheet is running. After changing the activator, the effect of this change on fiber level must be evaluated. The shift supervisors have already been checking this informally, but now they have a system to formally document their procedure.

If the fiber level is still abnormally high, the area supervisor will use the area response sheet. This response sheet will instruct the supervisor, for example, to change the type of cellulose being used in the system. The supervisor will also make other decisions, such as equipment overhaul or request changes requiring capital expenditure.

Project Results

The collecing tank fiber level for System 1 has been reduced from an average level of 205 ppm to an average of 134 ppm, and the System 2 fiber level has been reduced from 93 ppm to 74 ppm.

The PM system indicated that, because System 1 achieved four consecutive weeks at an average fiber level of 175 ppm, a reinforcement celebration should occur. The Film and Fiber Esters acetylation section

Table 5
Shift Supervisors
High Fiber Procedure

Date _____ Time _____ System _____ Blend No. _____
Fiber_____ Collecting Time_____ to_____

This is a response to:

____ 1. Run of 15 above target
____ 2. Run of 7 above UCL
____ 3. Run of 2 above 550

1. Call Maintenance to check the activator bearings.
2. Change activators. (1) From_____ to_____ (2) From_____ to_____
3. What was the first fiber grading after change?
4. After changing activators, is fiber grading out of control?
   a. Above upper control limit?   Y/N
   b. Run of 7 above target?   Y/N
5. If still out of control, go to Step No. 2. above.
6. IF STILL OUT OF CONTROL AFTER 2 ACTIVATOR CHANGES, NOTIFY FIBER ACETYLATION SUPERVISOR.

operators (which includes activation operators), the shift mechanics, the day mechanics, and electricians were invited to the celebration. During the celebration, the scoreboard and other charts were displayed while the area supervisor and department head explained to the people the behaviors and results that prompted the celebration. Immediately after the celebration, the team met to establish new goals for continual system improvement.

The resulting lower levels of fiber in ester dope have reduced the filtration costs in the manufacturing filter tow. This process improvement produced an annual cost savings of approximately $200,000. The reduced fiber levels also eliminated the need for in-process testing of fiber and filtration, thereby saving an additional $15,000 annually.

In summary, the team was able to tackle a tough problem by using all three elements of Quality Management: SPC, PM, and teamwork. The team brought a process into a statistical state of control and, in addition, had several breakthroughs that lowered the overall fiber levels. They developed a unique control system that incorporated a hierarchy of control and responsibilities. By accomplishing these feats, the team effort resulted in process improvements yielding annual cost savings of $215,000 without any capital expenditures.

SUMMARY

A key point of this project is that every person in the organization, from operators to the CEO, has quality responsibilities. These include producing the output right the first time and continual system improvements. Whether it's an operator making batches of chemicals or a manager designing a feedback system, both have a responsibility to do it right and then continually strive to improve it.

People at the operating level are primarily responsible for controlling and improving their production processes. This responsibility involves problem solving and contributing to team efforts to prevent problems from occurring. As the company continues to develop first-line supervisors and operators, they will assume more responsibility for system improvement. Lower and middle management levels also have responsibility for control, but their primary responsibilities are planning, executing, and reinforcing of quality improvement projects. Managers at these levels sometimes do not delegate work that is actually the responsibililty of those at lower levels of their organization. This results in too many highly detailed reports and meetings to explain upsets. Generally, far too much managerial time is spent reviewing historical data for the wrong reasons, rather than using it for identifying improvement opportunities so projects can be initiated.

Another responsibility of lower and middle level managers is reinforcing improvement efforts. As stated earlier, people typically do tasks they are reinforced for doing. This reinforcement can be manifested by attending team celebrations or praising a first-level supervisor personally for actively leading the effort. Personal reinforcement is extremely important, because first-level supervisors interact most often with the operators. These supervisors must be clearly perceived as "quality champions" by the people they manage.

The quality responsibilities of top management may be the most difficult of all to define. Top management exercises very little control over the manufacturing process, but they have great responsibility for leading the improvement effort and for strategic planning. Their role involves providing constancy of purpose, resources, reinforcement, quality organization, and, perhaps most importantly a corporate environment that fosters continual improvement.

One example of how top management can reinforce personnel is the President's Award at Tennessee Eastman Company. Quality improvement projects are scored against certain criteria, and the projects that score a preset number of points are recognized as outstanding projects. The president of the company personally presents the President's Award plaque to the members of the project team. The fiber control team received the President's Award in November 1986.

Once the improvement effort becomes more mature, top management should begin to assess its effectiveness and make improvements as necessary to ensure continued success. This evaluation requires top management's in-depth knowledge of the quality concepts and techniques of SPC, PM, and teamwork and top management's experience at doing quality improvement projects. Only then can top management fulfill their quality responsibilities.

Chapter Six

Xerox Process Qualification For Suppliers

Martin J. Madigan
Manufacturing Technical Specialist
Xerox Corporation

This chapter describes requirements for certifying a piecepart via process qualification and a new method of working with vendors. Unlike the literature and presentations telling American industry that they must implement process control, be innovative, apply statistics, use control charts, etc., Xerox demonstrates how these procedures can be accomplished. Xerox states exactly what a supplier must do and how to do it and, supports that instruction with training. Xerox offers a complete description of what suppliers must do to survive in the business enviroment in the next decade. Xerox does not instruct industry to implement process control and then retreat to an ivory tower, as so many companies do. The company promotes process control. Each supplier is given the tools and shown as often as necessary how to use the tools. This complete instructional package leads to quality products and economic success. The system that Xerox implements is used throughout the world, and all suppliers, regardless of location, use the same criteria and even the same forms, modified for the language of their country. This permits the use of supplier quality assurance engineers from local regions rather than from a particular Xerox plant. The requirements remain constant, and qualified products can be shipped to any Xerox facility in the world without inspection by Xerox. This concept of process qualification and, hence, part certification permits successful implementation of central commodities management, direct shipping, and just-in-time planning.

INTRODUCTION
Quality often is an elusive concept and, hence, is addressed only when troubles or complaints become excessive. Then it is asked: "What is the quality control/quality assurance department doing? How can they let such poor products ship? How could the receiving inspection department accept such poor incoming quality?" These are but a few typical questions raised

in times of trouble and they express a complete lack of understanding of the premises of statistical quality control. The question is not what "we" are doing but the tired old what "they" are doing. If you continue to ask this question, your corporation will perish. Xerox recognized in the late 1970s that the "they" concept was a failure. Xerox had learned that poor-quality products from a supplier meant poor quality and increased costs in the plant. The nucleus of the "team" image was in the works—Xerox didn't see it then, but it knew something dramatic had to happen if Xerox was to remain a leader in its field.

The old way of doing business was to receive goods from a supplier at some centralized receiving location or warehouse, sample the goods, send the samples to receiving inspection, test the samples, accept or reject the samples, then send the remainder of the lot to its final or intermediate destination (including the samples, if possible). If the samples are rejected, a decision must be made to return the lot, sort the lot (internally or at the supplier), scrap the lot, keep the lot, the famous "Use as is" because Xerox needed it to keep the line going, or some combination of those choices. All of these are offerings to acceptable quality level (AQL), average outgoing quality level (AOQL), and rejected quality level (RQL). These are used by most as a license to ship defective material. Many have heard: "But you only found 1.5% defective material and my AQL is 2.0%, so I'm o.k. on this lot, right?" This is a familiar situation, and buyers say "yes." They are buyers—quality control personnel must show them the fallacy via statistical process control (SPC) training. A 1.0% AQL will surely be met by the supplier, and it means a defect rate of 10,000 ppm—and the defective parts cost just as much as the 990,000 good ones. But how much does the repair, rework, etc., cost to eliminate these defects? Excessive handling and inventories for protection are examples of costly waste generated by such a system. Xerox wanted to ship directly from the supplier to where the parts were needed.

How could Xerox accomplish this mission? Companies told suppliers for eons that they must improve their quality. Many books are written about quality and some introduce readers to methods for quality improvement. Xerox is one of the latter—Xerox felt to simply mandate quality was not enough because Xerox had been doing that for years to no avail.

To meet the challenge of this mission, a "team" concept was born. A team must train to operate efficiently. This chapter explains how the team, Xerox and its suppliers, performs process qualifications and how the statistical training involved in these qualifications is accomplished.

Process qualification (PQ) is required for all parts used by Xerox in the manufacture and maintenance of its machines and their associated supplies, thus all suppliers must learn the basic tools of (SPC). How is the PQ requirement met? Xerox began by launching a major training effort. In 1980, a 14-point quality program was implemented, and one task was training. First the internal workforce was introduced to SPC, and then the supplier base was taught SPC via a worldwide series of seminars.

Quality forums were held—and are still being held—all over the world to teach SPC to the supplier-base executives. These forums were followed by a two-and-one-half-day session on statistics for executives' operations people. These seminars were then reinforced by in-plant assistance provided by Xerox supplier quality assurance engineers.

From this sequence, a formal SPC education evolved. The intent was for PQ to be the same worldwide. This PQ system consists of five sequential modules that must be completed. The key PQ elements are Figure 1. Training is provided as needed at each step. The following describes each element of the PQ system and highlights some of the major points needed for a successful PQ.

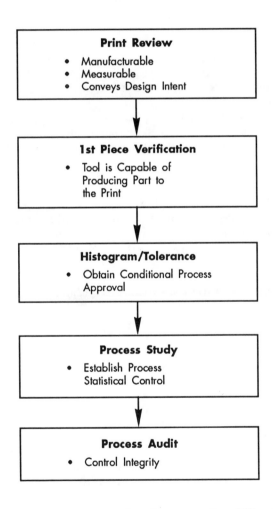

Figure 1. Sequence for Process Qualification

PRINT REVIEW

Print review is the first step in the PQ system. If Xerox does not have a correct drawing, then all is for naught—Xerox must first verify accuracy and completeness. Does the supplier need any of the required standards? Does the supplier understand the specifications and their intent? Can the supplier make the part? If so, can it be measured? Measuring some dimensions can be difficult and the supplier must state initially whether measurement is a problem. What about the requirements for gauging the part? Are special gauges and tools needed that can only be used for this job? If the gauges are special order, who will pay for them? These issues must be discussed, because a company and a supplier are a team, and both must exchange ideas and keep open minds. This initial stage is where planning for quality begins, and this is the start of early supplier involvement (ESI) process. Xerox involves suppliers, makes long-term commitments, and often awards contracts to our supplier base without bidding. If a supplier usually makes a particular shaft or motor, Xerox orders similar parts from that supplier and Xerox orders for the life of a program.

FIRST-PIECE VERIFICATION

First-piece verification has been in place, in one form or another, since mass production began. But has Xerox been doing it correctly? Xerox must use production tooling, not so-called soft tools (often still in use) or hard tools. Production tooling means using tools or processes that will be used for production runs—tools that will continue to be used as the program matures. A 100% layout is needed, and tool-controlled dimensions and the variables process dimensions must be identified at this time. The dimensions considered variables and that require process controls are those which will be studied further using statistical methods. Agreement on dimensions that require study will permit the vital few variables to be scrutinized and provide a start for the quality plan. Remember, this is just a first piece, and Xerox may change plans as process knowledge is gathered.

HISTOGRAM/TOLERANCE STUDY

Assuming successful progress through the print review and first-piece verification, Xerox does a simple feasibility study. The tool used by Xerox is the histogram, which is used in the form of a simple tally sheet. For the variables identified in the preceding modules, Xerox takes a 30-piece sample from a minimum production run. The production people on the floor, by assisting with the sample, are now members of the team. They can identify with Xerox, the part, and, with training, the end use of what they are producing. They can, with only a little training and practice, learn to perform all of the statistical work involved in PQ. The employees are essential, and not only must be involved in data collection, but they must also be allowed to make decisions. The operator—and only the operator—knows when a process is in trouble. The operator must be allowed to shut the process down, stop the line, etc. All production employees are vital to the sample run.

This run should represent what you would expect during a routine production run. A minimum run might be 100 or more pieces. The 30-piece sample is randomly selected so that it is representative of the run.

This sample is tested for the specified variable, and the data are recorded. The data are then plotted in tally sheet form and compared with the part specifications. At this point, several outcomes are possible, and decisions must be made. If the process centered at nominal and uses only a small portion of the specification, it is considered excellent. In this case, Xerox would proceed through the remaining modules to qualify the part. If the process exceeds the allowable limits, Xerox must shut it down and resolve the out-of-tolerance condition. Xerox must then repeat the feasibility study and any of the other modules deemed necessary. Finally, if the process is barely in conformance, Xerox calculates the estimated percentage of tolerance used and percentage out of specification. Again, as when the process exceeds acceptable limits, the process must be improved. Calculations in this case are based on the normal distribution and use of the "Z" statistic. Although this is a feasibility study, it can uncover serious problems that can be resolved with a minimum exposure to defective or even scrap parts. These three examples are only a sample of the many capabilities of a simple histogram. If the initial feasibility run was unsuccessful, it is repeated until the results meet the requirements set by Xerox.

PROCESS STUDY

The process study is the next module in the qualification cycle. When Xerox successfully completes the feasibility study, Xerox immediately runs a process study. If the feasibility data was handled dynamically, you might just continue the run. Another reason to get the operator involved. The purpose of the process study is to establish statistical control and describe the process visually. There are many guidelines and textbooks written on the "how to" of a process study, but Xerox provides a standard set of instructions. The word "study" was used rather than capability to avoid confusion. This study is not a capability study in the textbook sense but only an attempt to describe the process under routine operating conditions. If Xerox wants to know what level Xerox is on, what Xerox's average is, what range Xerox can expect to function within, or whether the process is stable (in statistical control?), Xerox can perform a process study. Xerox has standardized the process study methods. Initially, all studies are run on the variables previously identified. Data are collected in the prescribed manner from a normal production run, not a fine-tuned, special setup. Xerox requires 25 subgroups of five consecutive pieces. The subgroups are selected throughout the production run. This selection provides an adequate number of subgroups to use for plotting and evaluating the standard X-bar, and range charts can address stability over time—the histogram shape. Control charts and histograms are the foundation of PQ.

This evaluation indicates where Xerox currently stands and whether the process can provide acceptable parts now and in the future. Xerox uses estimates of sigma (R-bar/d_2) from the range chart and frequency distribution analysis, as needed. The range method of sigma estimate, assuming statistical control, is used to estimate the percentage of tolerance used or short-term process capability. The histogram results provide a sigma estimate used to calculate the "Z" statistic, process capability index (PCI), and, with the standard normal (z), estimates of percent defective. This study provides action statements for future production, which are an essential part of the piecepart quality plan. After a successful study, the plan can be completed.

The process study is the time to resolve any remaining problems. All necessary actions must be performed and problems resolved before a part can be certified. Usually this involves the team—supplier, Xerox, and the people. All groups, such as supplier quality assurance (SQA), design engineering, purchasing, manufacturing, etc., must be satisfied before problems can be considered resolved.

Problem resolution leads to a certified part. These parts are tracked by Xerox SQAEs via routine data analysis methods, such as control charts or histograms, as determined by the process study and noted in the piecepart quality plan. Certification is not taken lightly, because it permits shipping worldwide with no further inspection by Xerox. Certification is what permits central commodities management (CCM) to function successfully, and it allows a small supplier base. Certification permits direct shipments to distribution, manufacturing, or even to customers. It is a requirement for "just in time."

PROCESS AUDIT

A formal process audit supports certification and supplier controls. Parts are randomly selected from each supplier at frequencies related to their skill (PQ) level. In addition, Xerox has general guidelines that are used for all CCM and non-CCM suppliers. Before certification, SQA source verification is required for production shipments. A part may be qualified, but if the tool is not, then each shipment requires SQA approval until the tool is qualified. To aid in this approval Xerox established priorities such as quality sensitive (QS) parts, other program parts, other parts not yet certified, and parts certified via history, in that order. This system of priorities is driven by a requirement that all parts on new products will be certified before the start of production. Non-CCM suppliers are handled separately, and Xerox expects to always have a few, such as short-run shops, model shops, etc.

The goal of this process is to eliminate receiving inspection by the customer (Xerox). The logic is if the process qualification is successful, then there can be no defects, and inspection is unnecessary. The intent is to control the product as it is being made, make it correctly the first time, eliminate the waste of scrap and rework, ship to the customer the product requested and paid for, and produce Defect-free products. Unlike the slogans and ballyhoo of many quality control campaigns, Xerox asks for defect-free products, and then Xerox tells the supplier what must be done and how it must be done to obtain these products. Xerox uses SPC, and tells the supplier why at meetings in Xerox's facilities whre Xerox reports on progress and explains targets set by Fuji-Xerox, which has only 125 defective parts/million. Xerox met a target of 950 PPM by 1986 which was a 90%

reduction in five years, as Xerox began training in January of 1981. The goal for 1990 is 100 PPM.

MANAGEMENT PERSPECTIVE

Xerox will continue to measure defects by line fallout (LFO) in the production areas. Xerox found that using PQ has greatly reduced defects caused by measurement problems, and it has brought Xerox to a new awareness level. Xerox's prime defect problem is now related to attributes type defects. A Pareto analysis of LFO data showed that 60% of Xerox's defects fell into eight categories. Instead of wrong dimensions Xerox faced problems such as missed operations, wrong parts in the box, burrs/flash excess oil, rusty parts, packaging and handling problems, cracked parts, and defective threads. To address these problems Xerox needed to recognize that only the total quality control system would be effective. SPC is a part of TQC (Figure 2), but only a part of the whole system.

Management policies and attitudes about working with quality tools will lead to a successful team. Some processes Xerox will need forever, but for others, perhaps an $N = 2$ plan will suffice. $N = 2$ means that you take one piece at startup and, if it is sufficient, you run production. At shutdown

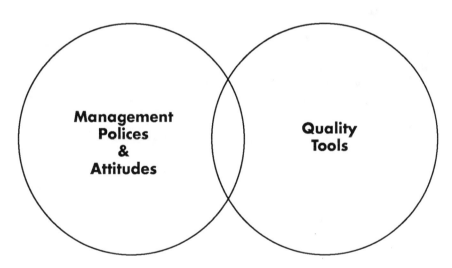

- Satisfy Customer Needs
- Obtain Goal of Perfection
- Hold Production Responsible for Quality
- Designate Everyone as an Inspector
- Control the Process
- Stop Production
- Improve Project by Project
- Involve Employees

- Employ Statistical Process Control
- Use n-2
- Implement Fail-Safe Operations
- Use Group Technology

Figure 2. Total Quality Control

you evaluate the last production piece and, if it is alright, you assume all parts are good—a PQ study can lead you to this plan.

Management must implement fail-safe procedures in the manufacturing process. A known, but usually ignored adage is, "Anything that is classified operator error must be assigned to management." Only management can control and, hence, change the system.

Management must project dedication to the customer, espouse perfection as an attainable goal, and recognize that production has sole responsibility for quality and that everyone is an inspector. Process control—not product control—is Xerox's target. Let people get involved, and let them stop the run when defects are about to occur. Only production workers know what they are producing. These concepts complement one another and lead to total quality control.

THE STAIRCASE TO SUPPLIER/CUSTOMER PROFITABILITY

Quality is the road to profits for the supplier and customer, or the staircase to profitability (Figure 3). The bottom step, data, is the beginning—don't put the data in a file cabinet. Let data lead you to profits. A simple thought process, a Pareto chart, is the next step. It can identify a variable that requires study or one that never needs to be studied again. Why waste test dollars on items that do not need testing? Some of the variables can be studied with simple control charts or histograms. Often these basic techniques or innovations will totally solve the problem or at worst will describe the process. Continue on to more complex issues using logic and fishbone (Ishikawa) diagrams, which when added to the previous steps may guide you to further analysis or may complete the problem solution. If further investigation is needed, then the avenue of advanced statistics is there. Designed experiments, regression analysis, Taguchi methods, etc., will aid in problem solving if you use them properly. Once you have finished

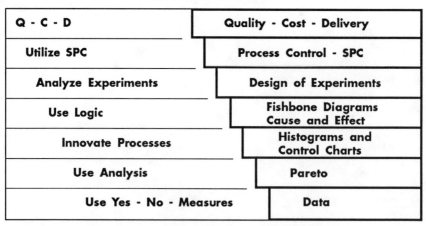

Figure 3. The Staircase to Profitability

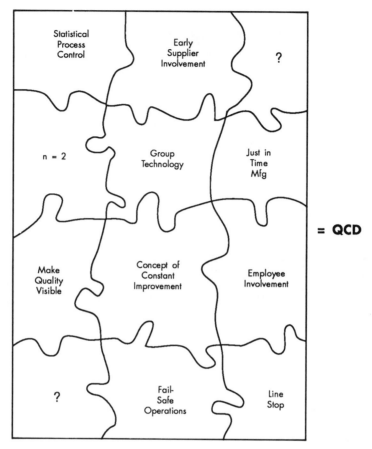

Figure 4. The Quality Puzzle

the experiment, study, etc., use the knowledge you have obtained. Implement SPC where indicated or n = 2 if that is the case, but implement intelligently. At the top step is quality, cost, and delivery. These are the drivers in today's business world. Quality is the generator that will drive costs down and, hence, price, permit on-time deliveries, and lead to increased demand. This is profit for Xerox and their suppliers.

CONCLUSION

In conclusion, Xerox is still learning about quality processes. However, Xerox turned the corner—Xerox is a team and is constantly seeking answers to the missing pieces in the quality puzzle, (Figure 4), and looking for ways to further improve quality. Total quality control, supplier training, employee involvement, and other methods are constantly being expanded and improved. It is Xerox's concept of never resting on its laurels, of never being satisfied with the status quo, and of constant improvement that will maintain Xerox's competitive edge in the ever-changing business world. Xerox—"Team Xerox"—can fill in the missing pieces of this puzzle, and then begin again.

© *The American Society for Quality Control, Inc. 1986. Reprinted by permission.*

Chapter Seven

Quality Improvement in the Pharmaceutical Industry

Thomas L. Fine
3M Riker

PHILOSOPHY OF QUALITY AND EXPECTATIONS IN THE ETHICAL PHARMACEUTICAL BUSINESS

The ethical pharmaceutical business (prescription drugs) represents $90 billion dollars in the marketplace in the world today. Unlike some industries for which quality may be a market advantage or a distinguishing difference from competition, quality is a prerequisite in the pharmaceutical business. Not just traditional product quality, but total quality in all aspects of 3M Riker operations. In the pharmaceutical industry, the term "quality" must be viewed in its broadest sense. 3M Riker is certainly concerned about the quality of its product, but perhaps more so than other manufacturing industries, 3M Riker is concerned about the quality of its processes, procedures, tests, and adherence to those tasks as 3M Riker makes the products which are eventually used by patients. To introduce quality improvement 3M Riker must first define what the term "quality" means.

At 3M Riker, quality means quite simply "conformance to requirements." This seemingly simple statement has a very complex meaning, because to understand requirements, 3M Riker must define them from many different perspectives. 3M Riker must also address quality improvement as anything and everything that contributes toward more efficient and effective business. Most importantly, quality must be defined from the customers's perspective, and therein lies the complexity.

3M Riker customers involve wholesale distributors, physicians, pharmacists, and, eventually, the patient or end user of its product. This broad customer base has defined requirements in many areas, specifically drug safety and efficacy, cost-effectiveness, service and availability, and, most importantly, confidence in 3M Riker's product. All of these attributes of total quality must be well defined, quantified, and met before 3M Riker can say that it has a product that meets its business requirements. The company's success or failure will depend on how well it understands these requirements as medical technology progresses and consumer demands change. As a result, 3M Riker spends a great deal of time trying to define these requirements and relate them to its operations in order to meet 3M Riker's ultimate goal of improving the quality of life of those people who use its products. Armed with a clear definition of requirements, 3M Riker can define and set business goals around results (outcome). The real management challenge lies in managing the performance (behavior) of people to attain those goals. In this chapter, 3M Riker intends to illustrate that quality improvement is an outcome attained in a systematic way through managing people's behavior in the workplace. The process can be greatly enhanced and done more effectively if approached scientifically by using performance management.

Designed-in Quality
Because the first precept of quality in 3M Riker's business has to do mainly with product quality, the company has to be concerned with product performance every minute from the time it is discovered, to the time it is eventually marketed, and to the time it is used by the patient. In today's pharmaceutical industry 3M Riker faces a very long lead time and significant investment to get a product to the marketplace. The Pharmaceutical Manufacturers Association's (PMA's) studies indicate that a pharmaceutical company today will spend, on the average, from eight to 10 years and $50 to $60 million dollars to bring a new molecule from discovery to the marketplace. The significant investment in time and money represents a very long and complex research and development path to define all aspects of a drug to meet regulatory requirements. In addition to meeting regulatory requirements, 3M Riker must meet the business requirements—that is, a proper return on the investment be made.

The process of designing in quality from discovery to market begins with the toxicology work to define a drug's safety for human use. All elements of use, both favorable and unfavorable, for human beings must be clearly defined in this phase of the investigation. Once safety is determined, 3M Riker must spend a great deal of time determining the efficacy, or actual usefulness of the drug, for treating illness or disease. This part of the designed-in quality process requires a tremendous amount of data collection, a great deal of research, and proof positive that the product, when used by a human being, will not only be safe, but efficacious in treating the particular malady. After efficacy and safety is determined, 3M Riker must then conduct a long and rigorous process to determine how to consistently and reliably manufacture the product and provide it in large quantities to the marketplace. This facet of the process is called scale-up/validation. This work is done to demonstrate precision, reliability, and repeatability of the processes 3M Riker uses to manufacture the product. As a responsible manufacturer, 3M Riker must complete this phase to comply with regulatory requirements as mandated by the Federal Food and Drug Administration (FDA). Very strict requirements must be met and proof positive must be produced at every step along the way. The validation process for manufacturing begins with the raw materials that are supplied, in some cases, from outside the company. Most pharmaceutical companies have a rigorous process for qualifying a raw material supplier. This not only ensures a consistent source of supply, but also a reliable source of supply in terms of the quality of the supplied product. Once raw material quality is defined and ensured, 3M Riker must have a clear definition of "work-in-process" controls. That is to say, every step of the process operation must be well defined in terms of testing, reliability, and repeatability. Proper controls in place during the process, ensures that a product will meet all of the finished product requirements as defined. These controls are supplemented by a long and arduous testing process of the finished product. This finished product testing not only ensures that the product meets the requirements when it comes off the production line, but also that it will meet "shelf life" requirements. 3M Riker must do lengthy stability studies to make sure its product's quality is maintained under stored conditions and it must define the useful life of the product, again, with regard to short-term and long-term safety and efficacy.

QUALITY MANAGEMENT IN THE WORKPLACE

Manage Behavior Not Outcomes

The key to success in meeting the requirements in 3M Riker's business is not only controlling the processes and ingredients that it uses to manufacture the product, but also managing people performance in all phases of the process. How 3M Riker manages performance is really the topic of this chapter, and the appropriate point to introduce performance management. If 3M Riker assumes that quality—or conformance to requirements—is essential in the manufacturing process for ethical pharmaceuticals, it must define what quality is and put the appropriate system in place to control and improve it. If it accepts that "quality," as conformance to requirements, is a standard that it defines, then 3M Riker must have a systematic approach for achieving that goal. In that perspective, "quality" now becomes an outcome, or a result.

The process by which 3M Riker attains quality depends on managing people's performance—people both inside and outside of its direct control as managers.

In today's manufacturing environment, performance management is one technology for achieving the goal of total "quality." For this discussion, performance management is defined as being "a scientific, data-based approach for managing human performance." 3M Riker carefully uses the term "technology" because it views performance management as an applied science—behavioral psychology. If 3M Riker applies the science of behavioral psychology in the workplace, the resultant technology is performance management. A successful manager in today's business environment should know the elemental concepts of behavioral psychology and, more importantly, should have a method for applying the principles in the workplace. The technology of performance management is based on four principles:

1. Pinpointing desired behavior;
2. Measuring performance of that behavior;

3. Delivering feedback to the performer on that behavior; and
4. Positively reinforcing the desired behavior.

If the desired behavior is positively reinforced, the likelihood of that behavior recurring is increased. 3M Riker believes, and the data demonstrate, that one of its most important jobs is to continually reinforce the desired behavior. This is a very easy theory to discuss and a very difficult task to accomplish. The company must first understand the distinguishable difference between behaviors (performances) and outcomes (results).

One of the goals (outcomes) of a department supervisor in the production area is to reduce the cost of department supplies, a challenge many supervisors and managers face in their normal work responsibilities. A manager might sit down with that supervisor, look at actual results, and set a goal for reducing that cost by a certain percentage. In their meeting, they define an outcome, or a level of result that is expected. Too often, however, the interactive process stops at this point. The supervisor is then left to figure out how to attain that result. Manager's jobs go beyond simply defining the outcome, managers must now work with the performer or, in this case, the supervisor to define the behavior the supervisor and the workers must adopt to achieve the desired outcome. 3M Riker usually arrives at this reduction by analyzing where the expenditures occur. Again, for the sake of discussion, 3M Riker determines with the supervisor, by analyzing his or her department costs, that the biggest department expense is hats. In discussions with the supervisor, and preferably even with the employees, pinpoint why hats are such a big expense. Maybe devise systems, policies, or practices that will reduce the use of hats in the workplace. Assuming through analysis with the supervisor and with the employees, that normal practice is to come in to work every morning and take a hat, discard the hat at the morning break, and take a new hat to resume work. During lunch break, that hat is thrown away, and after lunch a new hat is taken. During the afternoon break, the hat is thrown away, and after the break a new hat is taken. At the end of the day, the hat is again thrown away. Through working with the supervisor and employees, the use of hats can be reduced by saving and reusing the hats during breaks so that only once a day a new hat is taken rather than four times a day, as previously described. This conservation could reduce the use of hats fourfold throughout the department. Further, this result could be quantified in terms of dollars and cents. The outcome (reduced use of hats) is now pinpointed in terms of the worker's behavior (employees reuse hats) that can be measured, controlled, fedback, and reinforced.

Once this behavior is pinpointed devise a system of measurement—and maybe that measurement system would be a chart of collected data on the number of hats used per day. This chart should not only be displayed in the department, providing passive feedback, but should be discussed by the supervisor at the weekly crew meetings, thereby providing interactive feedback. If the supervisor engages in these behaviors as defined, focus on positively reinforcing that supervisor for doing this regardless of the outcome. In this case, the outcome should indicate immediate results—decreases in hat use and department expenses. Again, be sensitive to reinforcing the supervisor and the worker for engaging in the behaviors, not the outcome. This reinforcement establishes a means of providing more immediate reinforcement when the behavior occurs and not letting time pass before the outcome is measured. This is a very key point that must be grasped.

The importance of this reinforcment can be exemplified further. Assume that a company has a problem with tablet breakage in the bottle at the pharmacy and in the marketplace. This problem is not evident from finished testing before shipment from the factory to the distributor. Determine that the breakage problem correlates to improper hardness control during tablet production. Determine that if the operator runs a hardness test every 15 minutes and controls hardness at tablet production in accordance with defined control procedures, field-breakage problems are reduced by 90%. It may be determined that the most effective way of measuring results is by evaluating field-breakage problems. But, in fact, success or failure is directly tied to the operator's behavior in running the hardness test every 15 minutes and adjusting as necessary in the tablet operation. Reinforce that operator for running the tablet hardness test every 15 minutes, regardless of the current data on tablet field breakage. This practice ensures that the reinforcement the workers receive is more immediate to the performance and will help ensure the desired outcome of reduced field breakage, if correlation is correct. Without specifically discussing performance management as a technology, note one very important point. The most effective practice for employees, supervisors, and managers is focusing on behaviors and positively reinforcing them to produce the desired outcome. Very often people become complacent and totally ignore those people who are performing desired behaviors and, in effect, delivering the desired outcome. Companies focus on the problem performances and take for granted those employees who are doing what is expected. Of course, the reality is that today's manager must do both.

Improve effectiveness by paying more attention to those who do their jobs well and reinforcing their exemplary performances. Positive reinforcement takes many forms, but it must be something which is meaningful to the performer, such as the following:

1. a verbal praise or written letter of commendation;
2. recognition of a person in front of their work group or peers;
3. tangible reinforcers (jackets, caps, dinners, lunch, etc.); and
4. compensation increase tied to performance.

An effective way to understand what is meaningful to a person is to ask that person. Managers might ask subordinates at times to provide them with a "menu" of reinforcers they would like to receive. These range from memos to trips for the weekend. This is a useful tool for a boss to determine what kind of positive reinforcement is realistic and appropriate.

CASE STUDY

Two years ago, Mr. Alan F. Jacobson, 3M chairman and chief executive officer, defined three critical outcomes for every operating division of the company that he felt were necessary to achieve to assure continued success and business growth. These three outcomes also embody all elements of the total quality improvement process and track progress of total quality improvement implementation in measurable and consistent terms for the entire company. These three outcomes are the following:

1. labor cost reduction;
2. critical cost of quality; and
3. critical cycle time.

3M Riker immediately set about organizing various improvement programs under these three categories, and specifically focused on critical cycle time as the outcome to provide the greatest opportunity for business. This outcome involved the largest number of cross-functional teamwork activities to accomplish the improvement desired.

3M Riker first approached the project of reducing critical cycle time conventionally, thinking that the greatest improvement would come about through material control changes. 3M Riker established a task force (quality action team) consisting mainly of material control people to brainstorm and pinpoint some areas that could be changed to result in lower critical cycle times. Critical cycle time is the average time raw material is received in a facility until the time the finished product goes to the customer.

In a project, 1985 was designated the base period, thus data on actual experiences were readily available. The goal, or outcome, as defined by a chief executive officer was to improve results over the base period by 35% at the end of five years. At the outset, of course, 3M Riker felt this was a very challenging goal and one the company would have great difficulty in attaining. So, armed with complete understanding of the outcome and the challenge, a quality action team was formed, and the process of pinpointing began. As mentioned earlier, the quality action team consisted mostly of people from the material control department, thus most of the initial pinpointing was focused on activities and behaviors of material control people. As the pinpointing progressed, the company learned that not all of the behaviors pinpointed were within the realm of the control and material control analysts. Many of their activities depended upon outcomes from other departments and other activities that fed into the production planning/material control process. The pinpointing activities of the quality action team soon spread to include other functional departments, both inside and outside of manufacturing.

As specific behaviors and outcomes were defined, 3M Riker needed to include representatives from the accounting department, marketing department, data processing department, production, purchasing, quality assurance, etc. This led to a series of additional quality action teams in various areas to work on improving the performance of the material control department. Figure 1 illustrates the various subsets of behaviors and

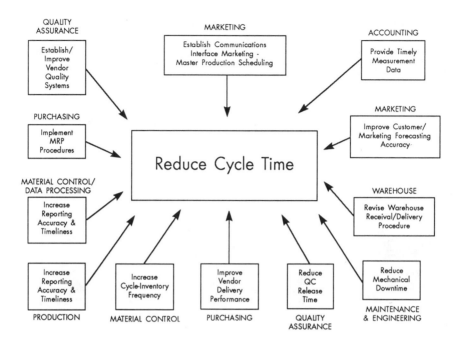

Figure 1. Quality Action Teams

outcomes that were identified, which all contributed to the overall goal of reducing critical cycle time.

Some type of structure to monitor and measure all of the various quality action team activities was needed. This led to a quality action team in each work area, reporting to a steering committee made up of key management personnel on site. The charter of the steering committee specifically was to review progress of each quality action team on a monthly basis and provide reinforcement for progress to the leaders and the members of the various quality action teams. All of these activities were viewed by the steering committee as one umbrella project of reduction in cycle time. 3M Riker chose this to focus on as a significant quality improvement project for its operation. 3M Riker's main concern at the time was that 3M Riker, as a management group, was able to devote enough time and attention to the individual quality action teams to ensure proper recognition and reinforcement to keep the activity moving forward. As the project progressed, 3M Riker's management group found this review and reinforcement activity to be very productive.

The company was constantly searching for new and creative ways to reinforce, both tangibly and intangibly, and soon the whole project became a lot of fun for everyone involved. To see certain supervisors displaying the kind of leadership ability in managing team activities that before they had not displayed was rewarding. The company observed various individuals in the workplace taking part enthusiastically and working more closely with their co-workers in their own department than ever before.

Another interesting "phenomena" was the amount of interdepartmental cooperation that took place to achieve certain outcomes. This spirit of cooperation, enthusiasm, and teamwork soon began to permeate the organization at all levels. One specific concern throughout this activity was to ensure reinforcement of various managers for working together. 3M Riker had initially emphasized to managers the behavior of working together more cooperatively. Often taken for granted in any organization is that key management people will work together effectively and cooperatively. In most organizations, however, this does not always occur. 3M Riker felt that if it could specifically identify outcomes which would require interdepartmental teamwork, and institute projects involving various managers and department heads, the company could then provide reinforcement to them for engaging in that behavior. They, in turn, could reinforce individuals within their respective organizations for the same behavior.

When the project really got rolling the company saw many of the classic "barriers" coming down. Barriers between marketing and manufacturing, between accounting and purchasing, and barriers between quality assurance and production. For the first time, many people were seeing the value in working together cooperatively toward a common goal, and in fact, were being reinforced by their management for doing so.

Figure 2 shows the measurement of 3M Riker's results so far. The results have exceeded even 3M Riker's expectations. The original goal, as defined by the company's chief executive officer, was to reduce cycle time 35% in five years. At the end of two years, cycle time had been reduced by over 50%. The new goal is to reduce that time by 60 by the end of year five. Throughout this project, 3M Riker has improved its customer service results and reduced its operating costs.

PAYBACK

The question asked any time quality improvement or performance management is discussed is: "What is the payback to the organization?" This case study illustrates several obvious paybacks, such as the following:

1. reduced inventory;
2. improved customer service;
3. cost-effectiveness; and
4. operational efficiency.

These outcomes are expected from any quality improvement process that is worth undertaking. In 3M Riker's particular case, it realized a "bottom line" savings in the millions of dollars. The company also realized, however, many more less obvious paybacks to the organization that it had not thought possible at the project's beginning. Everyone in the organization, for example, has gained a heightened knowledge of how 3M Riker's organization works. Through the pinpointing process 3M Riker has gained knowledge and understanding outside of its work areas that it never had before.

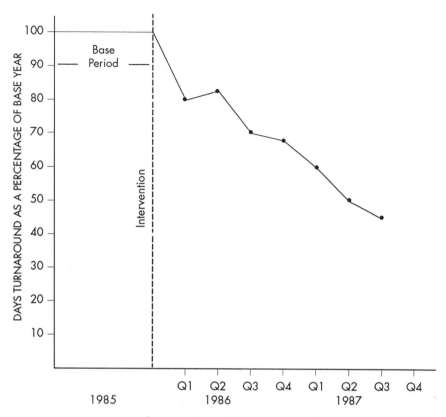

Figure 2. Critical Cycle Time

This knowledge has proved invaluable time and time again as employees have addressed certain problems in their day-to-day work and have worked together cross-functionally to solve them. Perhaps more importantly, there has been a great deal of self-realization in 3M Riker's organization, which has led to a heightened level of job satisfaction for all. There is no doubt that the quality of work life at 3M Riker has improved through this project, and the taste of success has created a growing appetite for more success. 3M Riker no longer has to ask its employees for ideas to improve its operation and now it is struggling as an organization with how to manage the influx of those ideas for quality improvement. Everyone in 3M Riker's organization has developed a sense of responsibility for managing their own work activities more efficiently and effectively. When employees are asked today about what they have gained from the quality improvement process, there are many different opinions and observations, but there are a few that can be highlighted that are common to everyone:

1. greater job satisfaction;
2. less chaos in day-to-day activity;

3. more time to focus on problems;
   4. a greater feeling of cooperation and enthusiasm toward the job; and
   5. a shared feeling of "ownership" toward achieving business growth—really making a contribution.

CONCLUSION

What 3M Riker has sought to illustrate, in the pharmaceutical business, as well as any business, is that there are certain requirements which must be met to attain "quality" performance. These requirements must be understood, well defined, and communicated to all of the people in an organization. The goals for quality improvement must be defined and measurable and perceived as an outcome of proper performance of individuals and groups. The specific behaviors of those individuals or groups must be defined (pinpointed). Performance must be measured, and management must ensure that feedback and positive reinforcement are delivered to the performers.

3M Riker has found that performance management is a very effective tool for implementing its integrated quality improvement process. Performance management is invaluable in providing incentive to keep quality work on-going and to give all individuals the feeling of satisfaction that comes with accomplishment. As long as quality improvement is viewed as the impetus that makes an operation more efficient and effective and, furthermore, is an outcome that results from proper human performance, then any business can be successful at total quality improvement.

Chapter Eight

Statistical Process Control Application To the Aluminum Extrusion and Drawn Tube Process

Robert F. Wolf
Quality Improvement Supervisor
Aluminum Company of America
Lafayette Works

OVERVIEW

This chapter includes elements of statistical process control (SPC) most applicable to the aluminum extrusion and drawn tube process. Included are the following key areas of discussion:

1. the unique challenge of applying SPC tools to the aluminum extrusion and drawn tube process;
2. the impact of initial training on the SPC process;
3. the use of the principles of quality costs and Pareto analysis to focus efforts;
4. the use of a flow process chart to further identify areas of SPC application;
5. the breakthrough achieved by applying the precontrol concept for process control and data collection; and
6. the gain in information analysis that was obtained by applying a computer software data package to precontrol information.

INTRODUCTION

This chapter addresses the SPC taking place in Alcoa's Lafayette Works. The Lafayette facility has a plant area of approximately 2,500,000-square feet and employs between 1,400 and 1,500 people. There are approximately 320 salaried employees. The hourly employees are represented by the Aluminum, Brick and Glass Workers' International Union. The plant is a fully integrated ingot, extrusion, and drawn tube facility with press sizes ranging from 900 to 15,500 tons and draw benches ranging up to 250,000 pounds. Ingot, extrusion, and drawn tube products are produced for markets including aerospace, transportation and defense, and general distributors of aluminum products.

This chapter will review the application of five of the seven tools of SPC to the aluminum extrusion and drawn tube process. These tools include: Pareto charts, flow process charts, histograms, run charts, and control charts. The tools of the fishbone or Ishakawa diagram and scatter diagrams are not included, however, these two tools have been used extensively at Alcoa's Lafayette Works. The fishbone diagram has been used in most of our team problem-solving efforts. The scatter diagram has been used in conjunction with capability studies and designed experiments.

For the extrusion and drawn tube process, applying the tools of SPC effectively presents a unique challenge. The product, as it is processed through the various workstations, is not identified in parameters a customer recognizes until it reaches the final operations. In the plant, the product is typically identified as logs, billets, extrusion charges, tube stock, and finish product pieces, whereas the customer relates primarily to the final product specifications, piece count, and packaging.

In addition, the extrusion and drawn tube process is not a continuous operation, but rather a series of discrete processes in which product variables change during each step. The planned lot sizes typically are small, with small press products (900 to 4,000T presses) averaging approximately 40 pieces. Large press extruded products (5,200 to 15,500T presses) average 10 pieces per lot, and drawn tube products average approximately 150 pieces per lot. This does not provide large amounts of data for applying the tools of SPC. Added to this problem is the complexity of the extruded shape, problems of ovality and eccentricity for tube, the factors of flatness and twist for extruded shapes, and a large variety of alloys.

Training
Complexity and diversity did not stop the Lafayette Works management from proceeding with the SPC effort. After conducting a plantwide awareness session for all employees, 10 hours of training on the seven tools of SPC were developed. The training was conducted in four two-and one-half-hour increments, with homework between classes. The sessions were conducted for all supervisors, technical support personnel, and selected hourly employees. The training was similar to that of most SPC-I type programs now available from many sources. Homework consisted of actual case studies of four successful SPC applications in the plant. Lafayette Works developed these applications before the training classes were conducted. These case studies included the following: 1) fishbone analysis and variables control charting of an ingot finishing operation; 2) a capability study of different methods of precision-sawing extruded forge stock; 3) control charting of ingot heating temperatures at the extrusion press; and 4) attributes charting of incoming production data sheets for the data processing operation. These case studies were the best received part of the training program. However, applying the knowledge learned from this training was minimal among employees.

There were many concerns about the time involved to collect and analyze data, some confusion about natural process limits versus specification limits, and some fear of the term "standard deviation." To institutionalize SPC, management had to display a continued commitment and personal interest to prevail over these fears.

Initial Applications

Initial attempts at applying control-charting techniques proved to be time-consuming, but the well-designed studies did yield some useful information. For example, a 6XXX series, alloy-capability study performed on a direct press indicated how the stretcher operation affected mean dimension and dimension variability. Figure 1 graphically illustrates the change in front-end and rear-end dimensions for each extruded charge of an extruded rectangular bar.

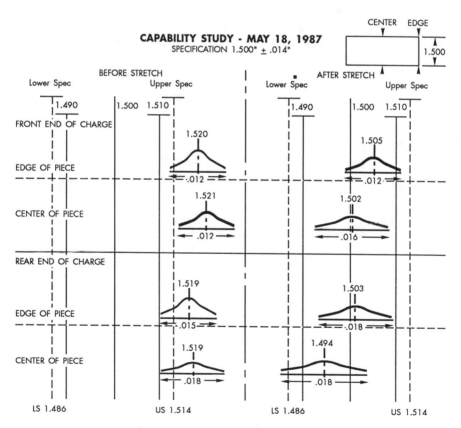

Figure 1. Extruded Rectangle

The information presented for data collected before stretch were very similar for means and distributions. The information for the data collected after stretch indicated there were significant changes, as follows:

1. the mean dimension for the center of the rectangle was reduced more than the edge;
2. the mean dimensions for the rear of the charge were reduced more than the front end of the charge; and
3. the magnitude of variation increased.

Based on these findings a better understanding of the impact of the stretcher operation on mean dimensions and variability can be obtained. By using this type of information it is also possible to focus SPC measurements on critical parts of the extrusion cross-section.

Cost of Quality and Pareto Analysis

To better focus resources, Lafayette Works started to analyze its processes in terms of quality cost and use of the Pareto principle. Lafayette Works did this to concentrate on major problem areas and to assign resources to its problems.

Figure 2 is an analysis of the drawn tube process. Developing this cost of quality analysis and establishing the Pareto chart did not involve any elaborate data collection techniques. This information was primarily derived by selecting a representative group of 120 production lots and using the available information on these lot data sheets to develop the information system. Surface defects comprised the major problem area—within that category, dents and scratches were the biggest problems. The hourly workforce was involved in initial team problem-solving efforts in this area. Good suggestions came forth immediately, which had a positive impact on the surface problem.

In addition, a C-chart was developed (Figure 3) to increase employee awareness of the surface quality problem. The operator at each workstation randomly selected a sample of five to 10 pieces per lot. The number and type of defects were counted by the operator, recorded on the chart, and plotted as indicated. This chart will be used in the future for team problem-solving sessions and as a means of measuring improvement.

Process Flowchart Analysis

In addition to the cost of quality and the Pareto analysis, a flow chart of the entire operation was developed to determine points of major variation in the process. A simplified version of this chart identifying 16 of these points is shown in Figure 4.

Actually, much more detailed charts have been developed in operations. For demonstration, this chart is used to identify some of the types of SPC applications that could be applied, such as the following:

1. periodic surveys of the customer's perception of quality and service performance can be developed and posted on Pareto charts;

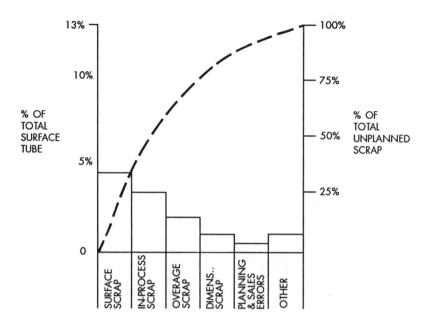

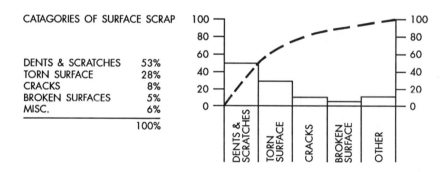

Figure 2. Pareto Chart - Drawn Tube Yield

2. average response time to satisfy sales and customer requests can be noted on a run chart or control chart;
3. average lead time and promise performance by product category can be monitored with a moving average type of chart;
4. computer control charting of critical alloying elements of individual alloys can be developed with exception techniques to indicate a high variability or out-of-control conditions;

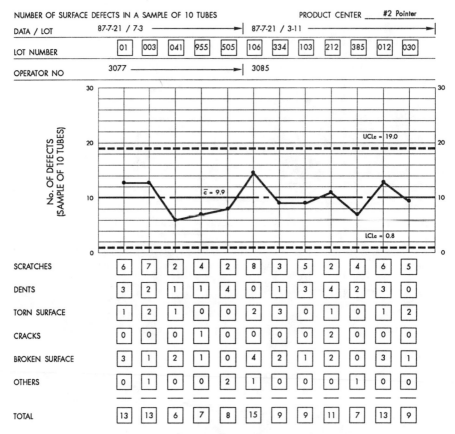

Figure 3. C-chart, Surface Defects

5. new die performance and average tool life can be charted;
6. ingot temperature, extrusion speed, and extrusion dimensions are variables that can be measured at the press and applied to analysis with control charts;
7. precontrol charting at stretch of front and rear end critical dimensions can be considered—however, more investigation is required regarding measurement techniques and equipment;
8. critical front and rear end section dimensions, length control, flatness, twist, and surface attributes can be charted at the saw operation through the precontrol techniques (discussed later);
9. tube stock dimension charting, surface condition attribute charting, and piece-weight control charting is practical through the pre-control techniques for tube products;
10. surface defect attribute charting can be performed at the pointing operation;

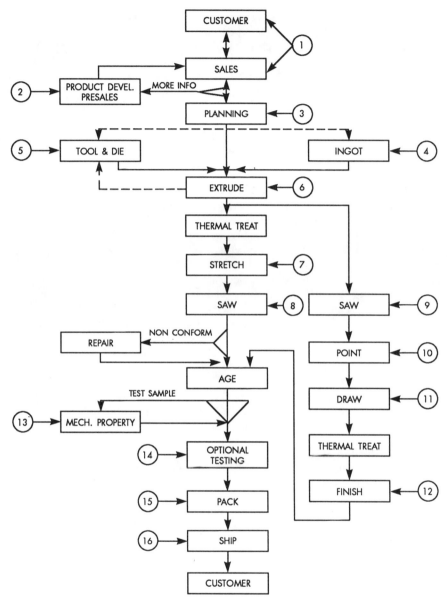

Figure 4. Simplified Flow Chart

11. product dimension precontrol charting is feasible at the drawing operation;
12. finished length control charting will fit the cut-to-length operation;
13. control charting and histogram analysis of mechanical properties can be maintained at the mechanical test facilities;
14. charting techniques, depending on specific special test requirements, can be considered where these tests are conducted;
15. overage and shortage analysis and paperwork error attribute charting is applicable to the packing operation; and
16. promise performance charting can be developed as a moving average type of chart as the product is shipped.

PRECONTROL CHARTING AT THE PRESS

Although X-BAR R charting proved to be time-consuming and impractical for small lot sizes, precontrol charting techniques proved to have some real merit. This concept cannot be accurately applied until the process displays a certain element of control, the typical product variables to be measured have displayed a normal distribution, and a capability exceeding the requirements of the customer's specification is indicated.

Assuming that the process and product meet these requirements, you will find the precontrol process is well suited to the extrusion and drawn tube process. In addition, employees relate well to this type of technique. Displays that highlight the major principles of the precontrol process are shown in Figures 5 and 6. Two of the charts that use these principles were developed for use at the 6XXX alloy series extrusion presses. They are shown in Figures 7 and 8.

Note in Figures 7 and 8 that the area colored red is outside of the specification and requires immediate action by the operator. The yellow areas that represent an area between plus or minus three standard deviations to plus or minus one-and-one-half standard deviations indicate a caution area. As indicated on the display charts this represents approximately 14% of the total population of a particular product variable. When one reading is obtained in the yellow area, the operator must immediately take a second reading. If there are two consecutive readings in the yellow area, the operator takes action regarding the process. If readings continually fall in the green area, which is one-and-one-half standard deviations in width, the process should be continued. Two positive factors regarding this technique are as follows: 1) the workers relate well to the stop light concept of red meaning stop, yellow meaning caution, green meaning go; and 2) no actual numbers need to be plotted, only points placed on a chart.

In Figure 8 measurements are indicated with a small "F" and "R." These letters indicate measurements from the front end and rear end of a specific extruded length. There is a significant difference between the front end and rear end readings. Thus, in the case of direct extrusions, the front and rear of an extruded charge can be considered different processes. Once these charts are completed, they are filed at the press. If a customer wants to review the data, it is available for that purpose.

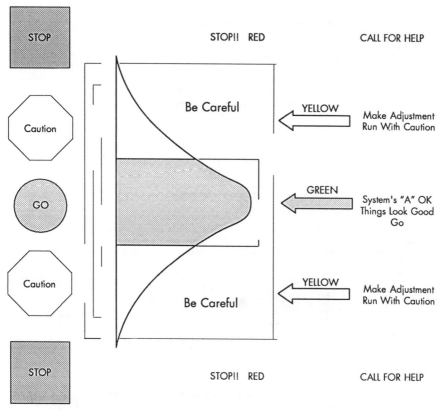

Figure 5. Process Control Utilizes the Statistical Normal Distribution

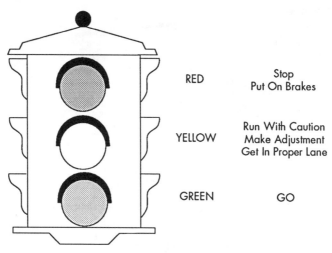

Figure 6. Process Control Traffic Light

95

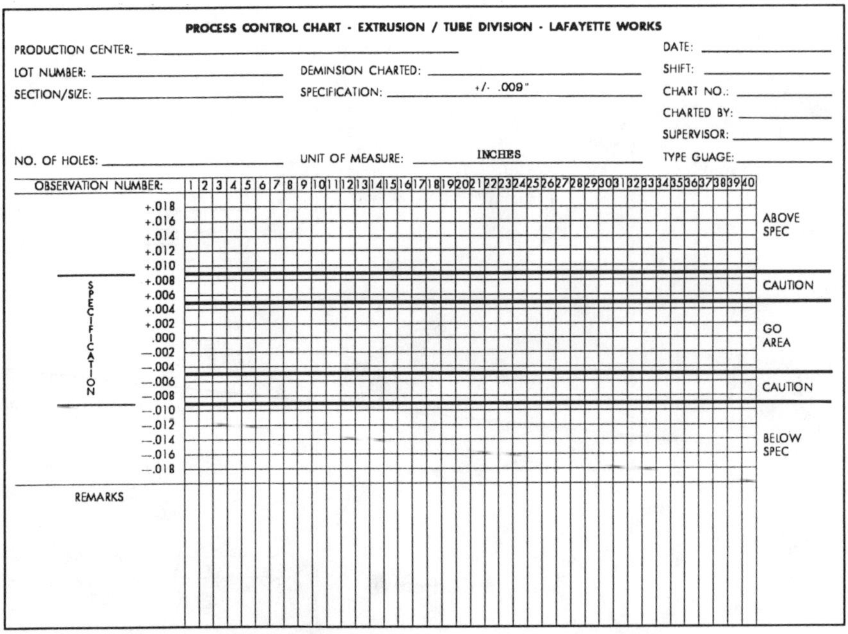

Figure 7. Process Control Chart, Blank

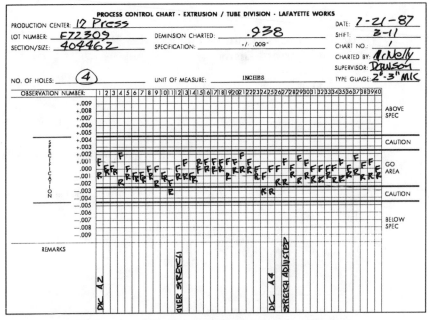

Figure 8. Process Control Chart, Filled in

Introduction of Computer Software to Precontrol Data

One of the weaknesses of the precontrol method of collecting statistical data was that it was difficult to use the data for further statistical analysis. Recently a software package marketed by PQ Systems called the SQC Pak, IBM Version 3.0 was purchased for application to precontrol data systems. Figures 9 and 10 indicate the type of control charts and statistical analysis information that can be developed from precontrol data. This has provided considerably more analysis power of the statistical data that is being collected at the press. Management can now relate to process control and calculate capability for use in both problem solving activities and discussion with customers. The software package is easy to operate and the data can be quickly converted from the floor charts to good statistical analysis information. This approach has served its purposes very well.

Precontrol as Applied to the Drawn Tube Process

While the 6XXX series alloy extrusion process allowed Lafayette Works to focus on one major place in the operation for data collection, the drawn tube process has a series of individual operations and dimensions change many times during the process. In order to get a good handle on drawn tube operations, a chart was developed which followed the product as it proceeded through the drawing and finishing processes. Figure 11 is a copy of a chart for one specific drawn tube specification that focuses very closely on eccentricity and ovality. Once again the precontrol colors of red,

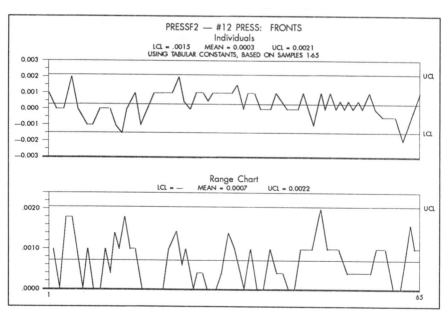

Figure 9. Control Charts

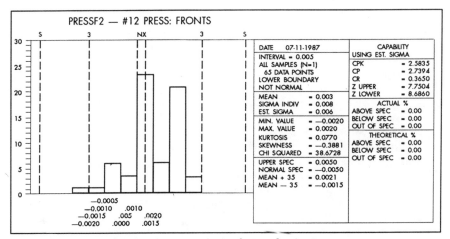

Figure 10. Statistical Analysis Data

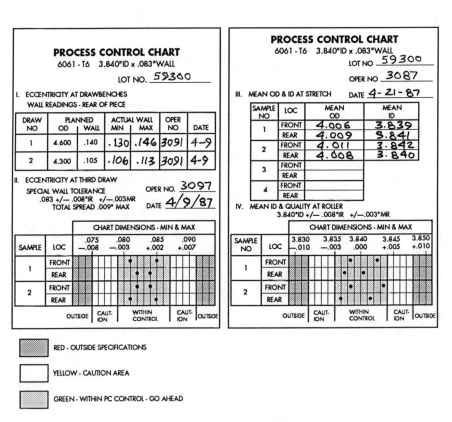

Figure 11. Drawn Tube Process Control Chart

yellow and green were used on this chart to highlight specifications. In addition, the SQC Pak was used to further analyze this data. Figures 12 and 13 are the resultant information displays represented as an individual moving average control chart and a statistical analysis display. Without this type of software package, completing the capability calculations would be time consuming.

CONCLUSION

Based on the application work successes that Lafayette Works has realized by developing the precontrol concept and adding a computer software package, the company plans to continue to use this type of application. The company will continue to use the team concept to develop charts and solve problems relating to charts and statistical analysis information. Workers at the Lafayette Works have shown a high level of receptiveness to SPC especially when involved in the SPC approach and problem solving activities. SPC training appears to yield the greatest impact when applied in conjunction with an actual manufacturing situation. The use of the precontrol concept and computer software analysis package has significantly enhanced this process.

Paper delivered at the International Aluminum Extrusion Technology Seminar April, 1988 and published in the Proceedings of the Seminar.

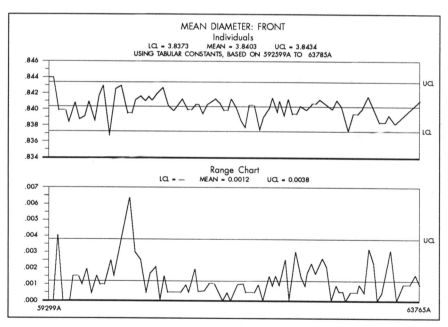

Figure 12. Drawn Tube and Individual Range Control Charts

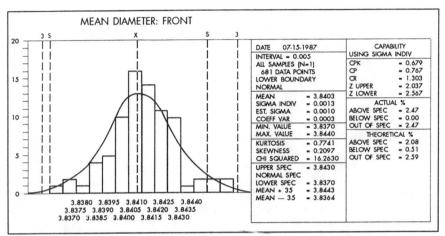

Figure 13. Drawn Tube Statistical Analysis Display

Part III. Management Tools

Chapter Nine
Applications of Quality Cost Concepts

William A. Golomski, P.E.
President
W. A. Golomski & Associates

INTRODUCTION

The quality cost concept with associated methods has become a very important part of many management systems throughout the world. It is used to identify opportunities to reduce cost, increase customer satisfaction, and track progress over time.

Any measurement system has strengths and weaknesses, and the quality cost system is no exception. (It doesn't measure business lost due to poor quality, but other reports do.) However, executives of large companies have learned when to use it with care. Quality cost systems are usually used as part of a quality management system with statistical methods as a centerpiece. Quality cost systems are increasing in use worldwide. Their use is primarily directed toward improving processes rather than products. This is based on cost studies of the comparison of product control versus process control. All processes are reviewed, including management processes.

Quality cost systems are used as an aid in setting priorities for quality improvement projects, studying cost trends to re-allocate resources, focusing multistep operations, measuring performance, and balancing efforts in reducing variation in design versus manufacturing. Measurements don't solve problems but can spur beneficial action.

Quality information can be obtained from a variety of sources. Some of it has greater impact when expressed in dollars. Robert Kaplan, of Carnegie-Mellon University, dramatizes the need for revitalizating accounting in an article entitled, "Yesterday's Accounting Undermines Production." He indicates that "direct quality indicators should be reported frequently at all levels of a manufacturing organization" (Kaplan 1984). Quality costs are such indicators. The four quality cost categories usually used are the following:

External Failure Costs. Cost associated with failures discovered outside the plant. These affect both cost and reputation.

Internal Failure Costs. Costs associated with failures discovered inside the plant.
Appraisal Costs. Costs of performing quality inspection and audits in-house as well as audits of suppliers.
Prevention Costs. Cost of quality systems, quality training, and education programs.

The detailed elements of these categories must be found in the chart of accounts or estimated.

External Failure Costs

- Product return, rework, or scrap costs, even if under warranty.
- Liability claims (distort the analysis if concentrating on day-to-day operations, but are important in estimating total quality costs of the firm).
- Customer complaints administration.
- Distribution and transportation losses. (Optional. May be a part of a quality cost system for the distribution department and the transportation department.)
- Cost of time of salespeople and others in resolving customer complaints (in some companies more than 25% of the sales budget and 35% of the R&D budget is in this category).

Internal Failure Costs (whether or not based on the standard cost)

- Scrap
- Rework
- Repair
- Downtime due to quality
- Retesting reworked or repaired product.

Appraisal Costs (whether done by people, robots, or on-line controllers)

- Inward goods inspection
- Tool & fixture inspection
- Process quality control
- Final inspection and test
- Instrument and on-line controller depreciation.

Prevention Costs

- Systems development & update
- Quality training and education
- Quality motivational programs
- Design reviews
- Reliability engineering and testing
- Pilot runs for process validation

- Participative problem solving
- Instrument calibration and maintenance
- Developing the strategic and annual quality plans
- Data analysis.

A quality cost system is a managerial tool for measuring quality. It is much more popular in the rest of the world than in Japan. Unfortunately, there is a feeling in many management circles that if the Japanese don't use it, it is no good!

Masso Kogure, of Tamagawa University, speculated on why there is this difference. He thinks it's because we don't have life-long employment and that managers and employees don't have the same degree of corporate loyalty as in Japan. Further, the emphasis in the United States on individual corporate herosim detracts from teamwork. Lastly, he surmises that U.S. quality professionals have more loyalty to their profession than to the company. This leads to searching for new ways to emphasize their importance. He feels that this is an attempt by quality assurance officers and advisers to get high visibility monthly (Kogure 1981).

In the United States, reducing the Cost of Poor Quality is viewed as the keystone of quality improvement by some organizations and as misleading by others. Nonetheless, these costs are only a part of total quality costs. Paracelsus, the great Swiss physician, early in the 16th Century made an observation that is useful in our discussion. He said, "There is no such thing as a poison; there is just a poisonous dosage" (Oxford University Press 1986). To paraphrase, "There is no such thing as a quality cost system being bad; it is only the misuse of the concept that is bad." Drinking water is good; drinking too much causes a depletion of electrolytes in the body and possibly, death.

We believe that quality cost systems are an important adjunct to the financial tools available to business and other organizations. Quality cost systems like any cost or control system, can be abused.

Standard Costs

During the 1920s engineers and managers asked the question, "How much more money would we make if everything in the plant always worked right?" Certainly this is an early quality cost concept and a good one. This was also a period during which cost accounting and cost engineering got their start. Unfortunately one of the main thrusts of cost accounting was to develop careful estimates of scrap and rework. This was to enable a build up of standard costs for a year. The emphasis was not on reducing them. Reports of actual versus budget were issued on a monthly basis. Today these reports are issued by shift and by production line in some operations. Stern questions are asked, such as, "Why is the material variance unfavorable?" The answer draws on a lifetime of creativity going back to one's childhood. To get someone off your back the emphasis is "fix the blame." Managements claim they are trying to search for the problem, but their actions after such interrogations show their true beliefs.

Any estimation problem has a statistical basis. The ideal situation is to get the process in a state of statistical control so stringent that only chance causes can disrupt the work. At this point there would not be any scrap or rework thus the material variance report would not be needed. Further, the diary or log kept on the chart would reveal the problem when it happens, if it happens.

Also, labor variance reports aren't needed if processes lend themselves to statistical control charts and the processes are in control.

Insurance and Health Services
In the 1940s Josef V. Talacko of Prague, Czechoslovakia, worked in a government unit in which he tried to use actuarial and statistical principles to predict costs. He was bothered by the unnecessary repetition of work in both administration and professional health delivery costs. There seemed to be no concern about losing business due to unsatisfactory service. Talecko tracked the costs within the department as well as their effects on other parts of the organization. These costs were accumulated by problem type, by department, and by office. He selected internal and external quality costs groups. He then conducted experiments on the effectiveness of various methods to prevent those costs. Talacko then applied some of these concepts on an informal basis with friends in production management at the Skoda automobile operations. Talacko's work was also applied in the United States in the early 1950s at United Airlines, Allen Bradley, and the Johns Hopkins Hospitals.

Manufacturing
There were a variety of early approaches to quality cost developed in the United States.

- Warren Purcell of Raytheon looked at basic work elements in work associated with manufacture, design, inspection, and other quality-related functions. These were costed out. They were then grouped into tasks and the tasks were grouped into jobs. Purcell had several interests—to find the cost of these various work elements, and to justify employment positions as well as how to reduce these costs.
- A. V. Feigenbaum, in his work, *Total Quality Control* (McGraw-Hill 1983), brought the concept of quality costs to General Electric and the quality profession in general. His emphasis was on the manufacturing system and its interfaces. J. M. Juran also mentioned quality costs in his first *Quality Control Handbook* (McGraw-Hill 1951).

Management Systems
The concept of quality costs is based on the belief that managements are intersted in having as few systems as possible to manage a firm. The systems are overlapping—one of them is the financial system, another the legal system, and another the human resources system. Obviously there are others. Each system addresses the concerns of the others in terms of its own interests.

In the 1960s quality was growing as an important strategic element in firms. When requests for budget changes for quality were considered, there usually was no sound basis on which to make a decision. Often the justification was, "It's for quality, it must be done." Another early problem was how to choose the quality improvement project that would have the greatest positive impact in the firm.

There were questions as to where to apply or use limited resources to improve quality on a complex production line. It is very difficult to convince Supervisor A to spend money to get a process to have less variation if the benefit shows up in Supervisor B's department. Quality costs can be used to convince a higher level manager that such scenarios should be resolved at his or her level.

No cost system alone improves anything. In fact, managerial behavior might focus on improving the numbers rather than solving the underlying problem.

Rationale For Quality Costs

1. Estimates the amount of money involved in poor quality and the amount spent to try to measure and prevent it. This is the total quality cost concept. When it is realized how large this amount is, a company is scared (or motivated) into doing something about it. Total quality costs such as 25% of sales are not unusual before quality cost improvement programs are implemented.
2. Justifies additions to staff and/or equipment when quality improvement is the main issue.
3. Helps to choose quality cost improvement projects based on projected return on investment.
4. Helps determine if a company is applying appropriate efforts in the various stages of a product life-cycle curve.
5. Determines measures of quality performance.
6. Provides data for annual and strategy quality improvement plans.
7. Measures the affect of the quality component in on-the-job training processes.

Applications

The following practical examples represent a broad spectrum of applications found throughout industry and government. The benefits from quality cost systems are identified in the following industry group case studies:

- Joseph Sesenbrenner, the mayor of Madison, Wisconsin, reported on the cost benefits of an employee involvement process in truck maintenance (MAQIN Conference 1987). The lowest bidder often delivered parts that failed more frequently, resulting in higher maintenance costs and higher total operating costs. Interestingly, the same information could have been captured earlier through a quality cost system. The question often is one of which system is the best to identify lapses and opportunities.

When customers of durable goods such as plant equipment, cars, or refrigerators want to evaluate what is best for them, they seldom have the information they need. Factors besides price, such as required maintenance and operator training are also important in determining the total cost to a customer throughout the life of a product or the period of ownership. This type of quality cost is called the life cycle cost.

There are a variety of practical issues of quality cost concepts. Even without defining the phrase "quality costs" most people would consider these questions important and based on quality:

- How is the cost of manufacturing related to raw material quality and end product reliability or stability?
- What is the manufacturing cost per unit?
- What is the quality cost per unit?
- What can cause testing time in assembly operations to go down?

Some reasons are obvious; some are not. T. J. Chartin in a 1970 article, analyzed it well when he said that improved fixtures, improved methods, and learning all have an initial effect. However, when testing, trouble shooting, repair, and retest time is some 40% of labor costs, there are new opportunities. This cost category is a good indicator of vendor and manufacturing process quality. Chartin showed how his cost system highlighted problems not generally apparent resulting in 66% reduced test labor costs in one year (American Society for Quality Control 1970).

- Yohei Koga used a flag diagram to indicate the part of a plant cost reduction target that was accepted by each of several subdepartments over the next 12 months. He also gave guidelines for relating selling price of life cycle costs (*ASQC Transactions* 1970).
- C. B. Rogers reported on the hidden costs of rework in terms of the effects on inventories, the production line, inspection, and customer satisfaction (*ASQC Transactions* 1971).
- W. N. Moore, of Westinghouse, gives the steps they take in managing quality cost improvement as finding the high cost areas, setting targets for improvement, planning action programs, organizing to take action, and measuring results (*ASQC Transactions* 1972).

The effort at Westinghouse is targeted for monitoring and encouraging the improvement efforts rather than with performance reporting on quality costs.
- Agnone, Brewer, and Caine of General Electric Aerospace observe that too many quality cost reports are directed at upper management while too few offer "insight" into specific problems that can be addressed effectively by first level management (*ASQC Transactions* 1973).
- John Hagan in a 1973 article reported that "hundreds of millions of dollars that may otherwise have been lost forever . . . came from the elimination of scrap, rework, and the reduction of clerical quality costs."

Many of his successes came in the computer data entry area (*ASQC Transactions* 1973).
- Robert Seamans, of the Air Force, as quoted by Kroeger of General Electric, said, "A failure that can be eliminated for one dollar during the concept formulation can cost $10 during the engineering development, $100 in production, and $1,000 in the operational phase (Institute of Electrical and Electronics Engineering 1971). At GE the cost of locating, removing, and repairing a failure and restoring a black box to acceptable performance under specification was calculated. At the lowest level of assembly the cost was $43, at the next unit level $154, and at the system level about $300. The goal was to drive failures upstream, even to the vendors' plants, and to eliminate them.
- J. E. Mayben, of the General Dynamics-Fort Worth Division, had done a magnificent job of showing the relationship of material review board approved rework hours as a percentage of total production hours versus the number of ships produced. The case study is an important example of learning theory concepts applied to a portion of quality costs. The percentages dropped from 40% to 15% from Ship 1 to Ship 2 and from 15% to 10% from Ship 2 to Ship 4. This is much better than one would expect. Usually as the units double, the costs reduce to 80% of the prior level (*ASQC Transactions* 1981).
- W. D. Goeller, of the Vought Corporation, has applied quality cost concepts to software quality assurance. He was concerned with the areas in which to prospect for gold. The cost for change in software maintenance was estimated as 30 times as large as that for changes at the design level. The test level changes were 7.5 times those of the design level (*ASQC Transactions* 1981).
- James Alback modeled the effect of various levels of prevention costs and appraisal costs on total quality costs based on experience. The rate of change of total quality costs was not constant. Even though small investments were made in appraisal and prevention, total quality costs went up. The investment wasn't large enough to have an impact. Finally, at the other extreme, additional large expenditures cost more than they were worth. Sometimes management tries pilot projects and does not get results. This might happen for two reasons: People don't believe that management really has a long-run commitment and the investment wasn't large enough to produce results. This effect was seen in some advertising campaigns for a new product (American Society for Quality Control 1975).
- David Seversky, of IBM, reported that the cost to remove and replace a component in the first stage of assembly was 46 cents in one application. The cost of finding, removing, and replacing that component in the field averaged $233.63 (American Society for Quality Control 1969).
- The City of Chicago was embarrassed and saddened when a bus and a car of teenagers were involved in an accident in which all the teenagers were killed. Subsequent investigations of busses indicated that many were without operating speedometers. The absence of a maintenance quality

assurance system was part of the problem. A quality cost system would have exposed that problem.

These examples should point out that quality cost systems can be of help in the following ways:

- quality performance measurement;
- economic trade-off decisions where quality is a factor;
- purchasing decisions;
- opportunities for improvement projects;
- a reporting system in dollars; and
- a measurement of importance to all operators, supervisors, professionals, managers, and executives.

Further examples of applications of quality cost systems can be seen in the following case studies from W. A. Golomski & Associates' files:

CASE XVI: Conglomerate, Inc.

A large and profitable subsidiary of a conglomerate, in capital goods, ran into trouble. Recently, their largest customer informed them that henceforth they would give 50% of their business to a competitor. The director of quality assurance called three consultants for help. He thought that an hour talk on quality costs and motivation would really fire up the plant managers.

The director of quality assurance said that they had as good a quality cost system as anywhere. It was patterned after Juran, Feigenbaum, and DOD. The signals coming out of it were increasingly bad. Internal failure costs were rising. The rework was so bad that products usually came out of the rework area in worse condition than whey they went in. Nonetheless, the director of quality assurance wanted help. When it was suggested that a 1-hour talk would not solve anything, he suggested a quick audit of his policies and procedures. Again, it was suggested that the problem was probably elsewhere. Finally it was agreed that a talk be given. As predicted, it didn't solve anything. Management got the feeling, especially the president and the executive vice president, that they were not using the quality cost reports properly. They felt that the reports, by themselves, would be sufficient motivators to encourage production managers to take action. Previous measures of motivation from the director of quality assurance were not effective.

After some reorganization of personnel at the director level it was decided that weekly meetings with the president should be held to discuss quality problems that needed attention. After three months, the internal failure costs went down, as did total quality costs. The weekly meetings were now held once a month. However, the production and quality assurance departments met weekly, or more often, if needed. This top management group took the necessary action to breathe life into a quality cost system.

CASE XVII: Electromotive Forces Company

Theodore Loo attended an advanced management training program at the Universal Transcendental Meditation Center. The program was an admixture of behavioral science and learning to raise one's level of consciousness. He came back as enthusiastic as he could be. As the owner of the company and principal inventor he had the freedom to try whatever he wished. Being very innovative, he conceived the idea of starting the day with 15 minutes of transcendental meditation (TM) for all factory workers and factory supervisors. He also decided to institute a training program for all people in the company. The feeling was that if everyone could improve their productivity by 1% the results would be fantastic. This would also complement Loo's concept of inner perfection (his answer to zero-defects). The out-of-pocket training costs would be $100,000.

The annual sales were $8,000,000 and the profit before taxes was $600,000. Loo thought that $100,000 was not a bad investment. He expected to get it back within the year, but he didn't. As a matter of fact, production went way down in most areas, and nonconformities remained at the same level or were slightly reduced in most areas of the Loo works.

Loo called a meeting of his key people in accounting, production, quality control, sales, and administrative services. He wanted to know why the sought-for goals were not being acheived. John Trosby, the production superintendent, said that it was obvious. Based on 400 employees, 40 hour work week, and an average weekly pay of $240 (including fringes) he showed the following chart:

$$.25 \text{ hr./day, or } 5/160 \text{ week devoted to TM}$$
$$+.75 \text{ hr./day, or } 15/160 \text{ week devoted to Training}$$
$$= 1 \text{ hr./day, or } 20/160 \text{ week lost from work}$$
50 working weeks per year
$$400 \,(20/160)\,(50)\,(\$240) = \$600,000 \text{ (total lost in \$)}$$
$$= \$100,000 \text{ (Consultant's Fee)}$$
$$= \$700,000 \text{ (Grand Total)}$$

Trosby said that the investment should be considered $700,000 at minimum. He pointed out that some orders were lost because of the reduced work week. The cost of lost orders should also be added in. At this, Teddy Loo grew pale. He said that this should have been called to his attention by the controller. But, he still wanted to know why even the $100,000 wasn't being recovered. No one seemed to have an answer.

Herman Pfrederick, the manager of quality control, said that he had been wondering about what part of the TM plus IP training he should charge to his training account in his quality cost system. It was obvious to him that the broad cost category was prevention costs. However, two things bothered him. One was that this brought his training account to a level way beyond that in the industry. Further, he had never been convinced of the value of the program. He went along with it because it was not charged to his budget. Later, however, he found out that the quality cost system

Management Tools

includes all quality costs, regardless of the department in which these costs lie. On that basis, and others, there was overspending for prevention.

Herman further suggested the use of the prioritizing principle, called by some the Pareto principle, to group problems according to their relative importance. He said that a few problems accounted for most of the losses—work on these first. Training or allocating staff or managerial time should be done on the basis of where you get the greatest dollar return.

CASE STUDY XVIII: The Seminole Pipe Company
The Seminole Pipe Company makes oil pipe. In recent years their business has had big swings in production. Their selling prices have been the lowest in the industry. This has given them a big advantage. Recently, however, they have been worried about it. When they have toured competitor's plants, their operations seem to be of equal efficiency, and seem to pay their employees about the same amount. Nonetheless, the influx of big orders for the Alaskan oil field project has them worried. They are concerned about the cost of field failures. These failures are discovered when the pipe is to be installed, or usually, immediately after field test. They have shipped pipe long distances in the past, but did not know if those shipments had a greater percentage of defectives than those shipped shorter distances.

The credit department handled all claims. They kept track of claims by salesmen and by major customer. There was a co-mingling of freight claims, overcharging, etc., with those associated with quality.

The company never considered external failure costs as a cost category. Consequently, these costs were not a factor in that they were not considered the responsibility of anyone. When claims came in, everyone possibly involved was called in to determine where to fix the blame and to watch things more carefully in the future. The final inspectors were told that if they did a top notch job they would end up with a promotion into management.

Management circulated pamphlets to their workers and hired a consultant to review the operation and make recommendations. The consultant offered the following suggestions:

1. The final inspectors are not interested in getting into management. Don't use promises that do not act as motivators.
2. Don't promote on the basis of a job well done. Promote on the basis of potential for success in the new job.
3. Establish external failure cost data for pricing and product liability purposes.
4. Utilize external failure cost data for pricing and product liability purposes.
5. Develop separate categories in the credit department for each category of claim, by product line, and distance or customer.

CASE XIX: J. H. Britain Company
The Food and Drug Administration has set guidelines for Good Manufacturing Practices (GMPs). As a part of this interest it was only natural to

suggest that process flow charts be developed and that critical control points (CCPs) be identified. The J. H. Britain Company has just had to recall 140,000 cases of infant food. They knew what caused the problem and it was easily corrected.

However, top management said that it would rather forego profits than have a public relations disaster like that occur again. The order came down from top management to foolproof every process. A process monitor has been installed between every production worker. Of course, there isn't room, but each process monitor takes his turn in standing between the production workers who have to move over to give him space in the line to observe. The surveillance is to last three months on each line for each product group.

During the second month of the program the monthly variance reports came in. Defectives are up, inspection costs are up, and the plant manager has rated only 82% on the plant managers' index.

John Jellicoe, the quality control vice-president, objected to the study initially, and still does. He said that the appraisal costs are going to increase in the long run. He said that it is unlikely that anyting new will turn up. In the meantime, the monitoring is considered a planning for prevention cost and the defectives are part of the appraisal costs.

Jellicoe pointed out that general management wanted to reduce the risk associated with the quality control effort. Jellicoe also pointed out that the balance of failure costs versus prevention and appraisal costs should be looked into. Simply overspending on appraisal costs might not be the best for the company.

CASE XX: Le Shan Industries

Donald Kinsella, the director of quality control for a can manufacturer, is one of the best known men in the field of service industries quality control.

At Le Shan Industries he is in charge of administrative quality control, as well as product quality control. He sees several opportunities in the administrative area for profit management. He'd like to work on them first because of the quick profit he could make. However, there are problems on the product side.

As a first step, he decided to establish a quality cost system covering both applications. He identified opportunities on each side and developed a single list of projects ranked in order of priority. These were distributed at a meeting of the quality improvement committee which he established.

He was surprised when he was shot down because his quality cost reports were too broad. Managers claimed that there should be two quality cost reports—one for product and one for services.

Kinsella reluctantly agreed to this. To do otherwise would disrupt the way the organization looks at products.

CASE XXI: Swedish Precision Products

Jon-Axel Steel, president of Swedish Precision Products, is a fantastic salesman. He has been able to attract good people to work in all departments. Some of the products sold were suggested to R&D by him.

Management Tools

Jon-Axel has been willing to try every new approach on his pet products. First it was work simplification, then value analysis and now quality costs. The idea was to try the concept out on one group of products and then extend these systems to others.

It has been eight years since a limited quality cost system has been in effect. Cristian Christopherson, the quality control superintendent, finds that he doesn't know what emphasis to apply to internal failure costs; external failure costs; appraisal costs; or prevention costs by product line.

He has made some cost estimates of the benefits of a broader program. Profits were down last year and are projected downwards this year. The timing is right, regardless of the politics. Narrow programs can't survive.

CASE XXII: Thomas-Cliffe, Ltd.

Leon Samson, of the Management Information Systems Department, believes in a quality cost system. He believes that it is a top-notch management tool to point out who is deficient in quality control. All that he feels should be reported out is total quality costs, and then these costs as a percent of sales and as a percent of manufactured costs. He believes that although the system captures internal failure costs, external failure costs, appraisal costs, and prevention costs by major product, these will not be reported out. Samson feels that the system's work can be justified on the basis of the aforementioned three reports. Whatever additional reports are requested will have to be justified by Zolten Tumay, the manager of quality control.

Tumay has an entirely different view. While he thinks the reports are useful, he thinks that more will come of a series of reports that can be a part of a quality improvement program. He is less interested in measuring the past than in improving the future. He would like to have quality improvement projects in each of four categories. He would also use the rule of thumb techniques that many have found useful in searching for the optimum on the quality cost curve.

Because each of the projects will have a different beneficial effect over time, he will use a discounted cash flow technique. Tumay is able to justify the more expanded report on the basis of plausible estimates of return on investment.

SUMMARY

In the preceding pages the concept of quality costs was introduced and cases XVI-XXII were selected from company files to show the following justification and/or reasons for implementing such a program:

XVI	Lack of top management understanding.
XVII	Over training.
	Autocratic management.
XVIII	External failure costs incomplete.
XIX	Appraisal was arbitrary.
XX	Program too broad.

XXI	Program too narrow.
	Management deity.
XXII	No quality improvement program built in.

In every case the opportunity for tangible profit improvement is present.

References

Alback, James. 1975. "Quality Cost." A presentation to the Kankakee Section of the American Society for Quality Control. American Society for Quality Control. Milwaukee, Wisconsin.

Chartin, T. J. 1970. Reducing manufacturing costs while maintaining liability and quality. *ASQC Transactions*. American Society for Quality Control. Milwaukee, Wisconsin.

Feigenbaum, A. V. 1983. *Total Quality Control, 3rd Ed.* McGraw-Hill Book Company. New York.

Goeller, W. D. 1981. The cost of software quality assurance. *ASQC Transactions*. American Society for Quality Control. Milwaukee, Wisconsin.

Hagan, John T. 1973. Quality cost at work. *ASQC Transactions*. American Society for Quality Control. Milwaukee, Wisconsin.

Juran, J. M. 1951. *Quality Control Handbook, 1st Ed.* McGraw-Hill Book Company. New York.

Kaplan, Robert S. 1984. "Yesterday's accounting undermines production." Graduate School of Industrial Management. Carnegie-Mellon University. Reprint #1114. Pittsburgh, Pennsylvania.

Koga, Yohei. 1970. Activities for reduction of user's cost. *ASQC Transactions*. American Society for Quality Control. Milwaukee, Wisconsin.

Kogure, Masso. 1981. Factors required for Japanese quality cost system. *ASQC Transactions*. Milwaukee, Wisconsin.

Mayben, J. E. 1981. Computer isolation of significant quality cost. *ASQC Transactions*. American Society for Quality Control. Milwaukee, Wisconsin.

Moore, W. N. 1972. Reducing quality cost. *ASQC Transactions*. American Society for Quality Control. Milwaukee, Wisconsin.

Paracelsus. 1986. *Oxford Companion To Medicine.* Oxford University Press. Oxford, England.

Rogers, C. B. 1971. Uncovering the hidden costs of defective material. *ASQC Transactions*. American Society for Quality Control. Milwaukee, Wisconsin.

Seamans, Robert, Jr. 1971. IEEE Reliability Symposium. Institute of Electrical and Electronics Engineers. New York.

Sensenbrenner, Joseph F. 1987. *Quality Improvement in Government and Governmental Enterprises*, (Audio tape). MAQIN Conference. Madison, Wisconsin.

Seversky, David. 1969. *Quality Cost.* IBM, Chicago Section. American Society for Quality Control. Chicago, Illinois.

Chapter Ten

Industrial Engineer Meet Dr. Deming: A Matter of Corporate Survival

Victor R. Dingus, P.E.
Technical Associate, Textile Fibers Division
Tennessee Eastman Company

INTRODUCTION

Today's professionals are declaring that management has a new job, that corporate survival is not a matter of buying new growth and consolidating assets. Survival is not based on fancy cost accounting, elaborate automated factories, or quick-fix quality meetings, circles, councils, etc. Corporate survival is based on the ability to deliver a high-value product to meet customer needs and satisfaction. The corporation must have a long-term strategy to guide it to improved fitness and long-term health.

What is the role of an industrial engineer? An industrial engineer must take a route like the following to ensure a company's survival:

1. read success stories of corporate survival;
2. identify how improvement is gained and maintained;
3. interpret and tailor improvement for local application;
4. identify an implementation strategy or process for intervening; and
5. get to work.

For those who are interested in corporate success, this chapter describes the what and how of continual improvement. Practical improvement discussed here is founded on Dr. W. Edward Deming's 14 points on business and protecting investors and jobs (Deming 1986). Other's ideas, thoughts, and practical applications are comingled with Deming's philoshpy for the key corporate issue of today: Quality management (and survival).

Industrial engineers have a choice: survival, suicide, or slow death. Deming's 14 points cover all of the currently identified technologies for which industrial engineers are stewards in modern corporate America. Many people believe that quality and production (productivity) are incompatible. This is a myth—if quality improves, then production (productivity) must improve due to less rework, less handling, less inventory, fewer assets, etc.

Improved quality and productivity are accomplished by those working within and on the system. Who is responsible for system improvement? Management, aided by staff resources—industrial engineers, statisticians, and other technical experts. Improvement must be a team effort by working continuously on the system and using data to assist in improvement.

Low quality means high cost no matter what automation is involved, what production volume is achieved, or what creative accounting is used. Many researchers have found that from 15% to 40% of the cost of goods sold is incurred for products not fit for use, for end of line sorting, or for quality appraisal. These costs can be reduced and can have an immediate bottom-line impact on the business. New equipment may not be the answer when old equipment may be capable (as statistically determined). Do not buy new equipment until you have gotten the most out of the old equipment. Statistical evidence should prove that the present process (system) is not reasonably utilized or capable before capital funds are approved for new processes and equipment. Measuring productivity does not improve productivity—measuring productivity is simply measuring a work process or a management system. The process must be changed to elicit a new response or measurement level. Change has to occur—the solution must comprise both statistical and behavioral thinking.

Historical Perspective
Deming's philosophy has an interesting history. He has known for the past 25 years that as the Japanese developed their skill in manufacturing (higher quality means lower cost), no American producer using conventional state-of-the-art approaches would be able to survive. Too many people believe that Deming merely teaches simple statistical quality control, and they miss the point. Deming has developed an entirely new concept of managing systems for machines and people. It is revolutionary and it works.

During World War II the statisticians of America, under leadership of such men as Juran, Shewhart, and Deming, pioneered new methods of control in the wartime industry. The result was America's ability to produce large quantities of high-quality armaments using an unskilled labor force (practically a miracle). These were not just methods of statistical analysis. They represented the beginnings of an entirely new way to look at factory operation. When the war ended, the mass markets of America were waiting to be satisfied, and skillful production management was not a necessity. By 1950 many lessons of the war were discarded. New managers came to run new factories. They perceived little need to increase quality. They did not study the new wartime principles, nor did the business schools teach them. Apparently, unsaturable markets swallowed goods of inferior quality. Americans accepted inadequate performance and compensated by repairing, reworking, or replacing. American management firmly believed that increasing product quality and reliability required an increase in cost, which was a serious misconception.

Japan faced a different situation. Their island, about the size of California with about half of the American population, faced a challenge. In 1950 the Japanese invited Deming to examine what they were doing. He studied their work force and became convinced his methods could apply. He held a meeting soon after with the top 45 industrialists, told them about his methods, and promised that within five years Japan would be an important factor in international trade. The industrialists had a limited objective—bring Japan back to pre-quality industrialization. Although many industrialists did not believe Deming, some tried his ideas. Without purchasing any new equipment, some managers reported improvements of as much as 30%. The industrialists began to compare notes. They realized Deming's method really worked, so they began to devote their time and energy to implementing his method.

CHALLENGE

What is Deming's method? It can be summarized in the following 14 points.

The challenge for all industrial engineers is to read, interpret, and customize the 14 points for local application and then to document and share success stories. The 14 points are presented along with ideas on how to implement them, examples of practical applications, and success stories.

Table 1
14 Points for Management

1. Create constancy of purpose toward improvement of product and service, with the aim to become competitive and to stay in business, and to provide jobs.
2. Adopt the new philosophy. We are in a new economic age. Western management must awaken to the challenge, must learn their responsibilities, and take on leadership for change.
3. Cease dependence on inspection to achieve quality. Eliminate the need for inspection on a mass basis by building quality into the product in the first place.
4. End the practice of awarding business on the basis of price tag. Instead, minimize total cost. Move toward a single supplier for any one item, on a long-term relationship of loyalty and trust.
5. Improve constantly and forever the system of production and service, to improve quality and productivity, and thus constantly decrease costs.
6. Institute training on the job.
7. Institute leadership . . . The aim of supervision should be to help people and machines and gadgets to do a better job. Supervision of management is in need of overhaul, as well as supervision of production workers.
8. Drive out fear, so that everyone may work effectively for the company.
9. Break down barriers between departments. People in research, design, sales, and production must work as a team, to foresee problems of production and in use that may be encountered with the product or service.
10. Eliminate slogans, exhortations, and targets for the work force asking for zero defects and new levels of productivity. Such exhortations only create adversarial relationships, as the bulk of the causes of low quality and low productivity belong to the system and thus lie beyond the power of the work force.

Table 1 *(Continued)*

11a.	Eliminate work standards (quotas) on the factory floor. Substitute leadership.
b.	Eliminate management by objective. Eliminate management by numbers, numerical goals. Substitute leadership.
12a.	Remove barriers that rob the hourly worker of his right to pride of workmanship. The responsibility of supervisor must be changed from sheer numbers to quality.
b.	Remove barriers that rob people in management and in engineering of their right to pride of workmanship. This means, inter alia, abolishment of the annual or merit rating and of management by objective.
13.	Institute a vigorous program of education and self-improvement.
14.	Put everybody in the company to work to accomplish the transformation. The transformation is everybody's job.

Reprinted from *Out of the Crisis* by W. Edwards Deming by permission of Massachusetts Institute of Technology and W. Edwards Deming. Published by Massachusetts Institute of Technology, Center for Advanced Engineering Study, Cambridge, MA 02139. © by W. Edwards Deming.

No single listing of ideas is an answer to all situations—however, any situation can be improved by applying at least one or more of these principles.

Point 1. Create constancy of purpose toward improvement of product and service, with the aim to become competitive and to stay in business, and to provide jobs. Purpose means improvement, and continuous improvement is a way of life. The new philosophy requires reevaluating the organization's purpose, the employees' obligations, and, especially, the managers' relationships with employees. To be competitive in world markets, management must satisfy a wide variety of people and view the corporate body from different perspectives. This can only happen through the following:

1) Management must innovate for the future, thus assuming faith in the future. They must allocate resources for new products, services, materials, methods of production, necessary skills, training and retraining, production performance, and user satisfaction. The innovation foundation is built on omnipotent successes of quality and productivity improvement. People can't expect the best ideas if skepticism exists.
2) Management must allocate resources for research and education to continually improve product design. Considering the next person in line as a customer or consumer is a primary strategy. The next person in the production process is the most important part of your production process. It is a mistake to produce efficiently, employing statistical methods, if everyone performs with devotion while delivering the wrong product.
3) Management should invest resources in facilities, maintenance, and reliability. Ignoring these areas can lead to producing unfit products for use by the customer or consumer.

As an internal consultant, the industrial engineer can promote corporate constancy of purpose by helping marketing organizations develop a survey or rating tool to assess customer satisfaction. Measures and statistics can be used to show progress and improve problem solving. Using team and consensus building, a survey tool can be created to focus on product quality, information exchange, service, and relationship. Even the customer can share in customizing the survey. This survey and the resultant team building yield a constancy of purpose that is data-driven, helps solve problems, and recognizes the people who constantly improve customer satisfaction. The course set by this tool provides for business tomorrow.

Point 2. We are in a new economic age. Western management must awaken to the challenge, must learn their responsibilities, and take on leadership for change. What's wrong with the old philosophy that dictates people should produce more, set quotas, measure productivity, and reduce cost using value-engineering methodologies?

The new economic age stipulates that improving quality results in less cost, not more. Why does productivity increase when quality is improved? Blue-collar workers can supply the answer—less rework, less handling, smoother production runs, fewer process changes, and a sense of pride in workmanship. Further, the customer can see the quality difference and may even order more products. A team using classical statistical process control methodologies of control charting, team-derived control strategies, and response sheets to document out-of-control situations for analysis, problem solving, and process improvements recently experienced greater customer satisfaction and purchasing volume.

The new philosophy considers defects unacceptable. These nonconforming parts can be statistically tracked and effectively scrutinized through problem-solving tools, which will lead to consistent production procedures. The people in the system can use these tools to operate effectively and to produce products to meet customer needs, thus alleviating future problems and establishing leadership in the product line.

For example, an industrial engineer and a production supervisor worked together, held team meetings with operators to identify consistent operating procedures, and developed both a control strategy for handling out-of-control points and a response sheet to document action taken. The result was a best-ever production run and the customer's order for an additional year's supply of the product.

It is time to change management's philosophy—a company will not survive in the new age by just meeting competition. Businesses must satisfy the customers' perception of their current and anticipated needs.

Point 3. Cease dependence on inspection to achieve quality. Eliminate the need for inspection on a mass basis by building quality into the product in the first place. No amount of inspection will force quality into a product. Inspection assumes there are defects and is not consistent with the new-age thinking. Inspection only increases the cost of goods sold. Though inspection might render improved outgoing quality, it adds cost. Mass inspection often overlooks some bad products and rejects some good products.

Another inspection/in-process control technique is to automate process controls based on engineering specifications. Statistical analysis yields product measurements that are uniform between specification limits. If people use classical statistical process control techniques in conjunction with group problem solving to change the process, improved process capability will result.

How does the industrial engineer address such a problem? Because of a recent customer complaint, an industrial engineer was asked to look at variability among inspectors. While this problem was addressed, the entire process was also analyzed by process flow charts, statistical analysis, Pareto diagrams, and team-problem solving. The study, found that the inspectors filled out citations whenever an off-quality product was identified. These citations were sent to the manufacturing section where they were used as audit trails to the machine and operator. These complaint citations provided the basis for setting quotas, rating the machine operator's performance, and occasionally adjusting or repairing machines. This was not effective.

The solution was a cooperative effort between the departments to limit mass inspection by instituting self-inspection and greater machine control with only random audits. The results were dramatic. Machine-control methods were improved, operator handling was improved via methods work and retraining, causes of defectives were found, and process changes were made. The citation sheets were discontinued, and operator feedback and recognition were implemented. Pride in workmanship was regained. Quality is up, rework is down, there is less inspection, costs are down, there is greater on-line, real-time control, more pride in the job, and happier customers.

Point 4. End the practice of awarding business on the basis of price tag. Instead, minimize total cost. Move toward a single supplier for any one item, on a long-term relationship of loyalty and trust. Price has no meaning without product quality. Thus, the purchasing agent and buyer have new jobs—to learn about statistical process control techniques and speak a common language with manufacturing. Why? How is a person going to meet customers' needs at a price the customer is willing to pay? Long term relationships should be established with suppliers who consistently meet your needs and expectations and strive to improve their ability to do so over the duration of your relationship. In effect, you are buying the supplier's process as well as the product.

Additionally, to meet customers' needs consistently, the number of suppliers used should be limited, including in-house suppliers. Understanding the concepts of process variability results in this conclusion.

Implementing this principle is difficult. People must understand raw material fitness-for-use requirements and accurately convey and negotiate these with a limited number of vendors and suppliers. People must enlighten vendors and suppliers to the new economic age thinking. Show them how to implement statistical process control techniques and provide for continued improvement.

Many companies, such as Ford, are adopting this new doctrine. Industrial engineers can assist purchasing in understanding statistical process control techniques and team problem-solving tools. This understanding can help build supplier rating systems that analyze price, service, quality, relationship, continual improvement, financial health, and labor climate. All of these factors interact to produce value. Eventually, a certification program can emerge as a business strategic plan and process. These partnerships resulting in process qualification and vendor certification should lead to very competitive positions for both parties. People cannot afford to ignore this strategy.

Point 5. Improve constantly and forever the system of production and service, to improve quality and productivity, and thus constantly decrease costs. Continued improvement is a basis of the Deming cycle and includes the following:

1. Recognize the opportunity.
2. Test the theory to achieve the opportunity.
3. Observe the test results.
4. Act on the opportunity.

Improvement is spiraling toward a customer's operationally defined need. The improvement represents one of a vital few from a Pareto listing based on your analysis, customer input, and anticipated needs. Once one project improvement is achieved, the team moves on to another vital issue. Note that in all operations there is potential for improvement. Each step in the product process or service can be improved. How?

Important to implementing the new economic era philosophy is understanding the need for products or services to fall within specifications. The old thinking results in a belief that all products and services are good—they are fit to use. The new thinking reveals that reducing variability of products and services yields the highest quality at the lowest cost. People need to recognize that economic loss occurs from any deviation from a target value. The loss may be minor for slight deviations, but grows as deviation from the target area increases. The loss may also be minor for this process step, but multiplies as you continue. Some losses are quantifiable and recognizable and some are not easily defined.

An outstanding example of an industrial engineer's assistance in continual improvement was evidenced at a textile manufacturing plant. In this situation, quality had been declining, and a rigorous quality inspection program had been initiated. This program in turn led to reduced productivity. The industrial engineer took a different approach—he used statistical process control, teamwork, and performance management technologies in concert to identify customer requirements, determine process capabilities, identify opportunities, implement improvements in all aspects of the process, and systematically recognize the people-improvement efforts. The results were approximately 100% improvement in productivity over a few months with small capital investment. In addition, the customer now rates this supplier as the best and brags about product quality.

Point 6. Institute training on the job. Changing company systems alone does not ensure continual improvement. A cornerstone in building competent management is to communicate expectations and job accountabilities clearly. Once these are identified, people must train managers in the quality tools and technology needed to change the way people think about and relate to the company. Companies must change the way they think about each other, change the way they solve problems, and achieve greater consistency in their work.

Training intensifies a person's knowledge. It must begin at the top—anything else will fail. After training, management must permit the knowledgeable person to help make changes for improvement. Management must remove barriers to applying the new quality tools and technologies of the new economic era. Having a strategy in place for using new-found quality tools and statistical thinking will greatly assist in breaking down many barriers.

Training should include using statistical thinking (control charts, process diagram charts, cause-and-effect diagrams, histograms, Pareto diagrams, scatter diagrams, check/response sheets, and design of experiments), behavioral thinking (understanding that people are rational and what they do is a result of either anticipated or known consequences), and operational methods training (show and tell, try and discuss, do and critique, and self-feedback). This three-pronged approach yields consistency, pride in workmanship, and continual improvement of job methods and systems.

For years industrial engineers have been training workers in routine use of their technology and tools. This training has usually been need-driven for a particular project result. This is the corporate strategy to survive. Previous experiences can be transferred to this new strategy, incorporating training expertise into the solution. Again, training must start at the top and cascade downward.

Management must get involved and lead the training, and they must be models of the new thinking era. Training experts, quality specialists, and industrial engineers must implement the strategy. This effort is called training with a strategy. New knowledge must be conveyed, barriers must be removed, examples must be set, team work must be initiated, and results must be shared. Training without a strategy and top management involvement will fail—these are the first barriers to remove.

Point 7. Institute leadership. The aim of supervision should be to help people and machines and gadgets to do a better job. Supervision of management is in need of overhaul, as well as supervision of production workers. A key concept is instituting leadership. As mentioned previously, this means model, don't dictate, the new way.

Management is more than a judge, naysayer, and pronouncer. Management should be a coach, teacher, enthusiast, cheerleader, helper, and counselor. The new job is to get everyone working in the system to help work on the system. This is management's responsibility and new job. How is this accomplished?

The answer is by statistical thinking and behavioral thinking. Juran says there are three things people must know for self-control: what to do, how they are doing, and what are the means to change. What is missing? Some believe it is the desire or will to change. Quality specialists talk about increasing will, desire, or commitment to change. Deming says that recognition or reinforcement is essential in the improvement process. Aubrey Daniels has shown that the best approach is promoting recognition and reinforcement by removing consequences in the system that provoke adverse behaviors and adding positive consequences to the system for what is needed. The science is called performance management (Daniels and Rosen 1983).

Industrial engineers are beginning to use performance management to generate desire and enthusiasm for and commitment to statistical thinking. Performance management and statistical thinking are necessary in understanding the variation of the process as controlled by people. Industrial engineers have been extremely successful at a number of manufacturing settings in assisting management with using statistical thinking, removing barriers, and adding positive consequences. When these tasks are accomplished, work becomes worthy and fun. The new theory of self-control asks what to do, how a company is doing, what are the means to change, and what is meaningful in the work place.

Point 8. Drive out fear, so that everyone may work effectively for the company. Without mutual respect, no statistically based management system, or any other system, will work. Measurement has been and still is used for more and better punishment. Quality thinking has been around for years, but has often died because many misunderstood it and feared its use. Management must now advance from the "I got you!" to a "How can I help you?" philosophy. Measurement should be—or rather must be—used for recognizing positive behaviors and for problem solving. Understanding and applying the basic law of behavior (i.e., behavior is a function of consequences) "will lead to driving out fear." Understanding this behavioral theory makes good management a science rather than an accident. Management becomes a systematic, conscious, and fun way to learn, initiate projects, change company culture, and improve performance. The recent book, *In Search of Excellence* (Peters and Waterman 1982) declares: "The excellent companies seem not only to know the value of positive reinforcement but how to manage it as well . . . . The systems in excellent companies are not only designed to produce lots of winners, they are constructed to celebrate the winning once it occurs."

Industrial engineers are helping all levels of employees to drive out fear by using a participative approach to problem solving and instituting new control methods. Everyone's ideas are solicited, and the best ideas are identified, tested, and monitored so that the productive activities being accomplished can be recognized. The activities are positively reinforced as well as the results of the project via a team celebration. These celebrations call attention to the results, the actions taken, those who participated, the impact and importance of the improvement, and the manager's appreciation of a job well done—similar to team sports. The assignments are made, the game plan is well laid out, the scoreboard is lit, and team members do their jobs and help others to get their jobs done. Finally, recognition is provided by the coaches. Thus, the main point is to drive out fear by using participative approaches, manage the consequences, and stick to the game plan.

Point 9. Break down barriers between departments. People in research, design, sales, and production must work as a team, to foresee problems of production and in use that may be encountered with the product or service. Viewing the total organization as diverse departments operating independently can devastate the company mission. Understanding that management operates processes and these processes interact, we begin to uncover the need for breaking down barriers between functions. Each function and its transformational process has a customer, whether it's an internal, external, or governmental customer.

Industrial engineers are assisting management in defining processes, flow charting the activities, and showing how these transformational processes relate to one another. These efforts reveal the need to simplify and remove barriers to achieve excellence in meeting the company's mission and customer needs. The flow-charting techniques have been demonstrated by Myron Tribus in a paper entitled "Creating The Quality Service Company" (Massachusetts Institute of Technology 1984). Tribus uses classic conventions to show process steps, meetings, decision points, and cooperation. Additionally, the functions and key positions are shown at the top of each flow chart's page. Blocks can be viewed by using the same charting techniques to show the individual block in detail.

Improvements in the functional processes are revealed as the processes are charted. These pictures show the barriers and provide answers to why we may be suboptimizing our mission and customer satisfaction. Also, management can improve the definition and measurement of internal products. Many improvement opportunities that can be taken care of within one department or function will be found. However, in many cases, the improvement opportunity will cut across functions. Multifunctional groups can best attack the improvement. Industrial engineers using classic flow-charting techniques on management processes can and have assisted in producing significant bottom-line results.

Point 10. Eliminate slogans, exhortations, and targets for the work force asking for zero defects and new levels of productivity. Such exhortations only create adversarial relationships, as the bulk of the causes of low quality and low productivity belong to the system and, thus, lie beyond the power of the work force. Slogans, production quotas, and zero defects are all limiting and do not alone produce the results the well-intended manager is seeking. At best, a manager gets compliance to these incentives. What are people seeking? The new economic era demands a different strategy—continual improvement, clarity of expectations, and pairing of incentives with meaningful consequences to sustain the new strategy.

A vivid example is an industrial engineer who was assisting a management group responsible for a batch production process. Performance standards had been placed on how many batches a crew was to produce each day. A long history of data showed that reworking batches was a frequent occurrence. Hence, quotas were met by the crews, but overall department output was low. The consequences in the quota system were favorable to reworking rather than doing it correctly the first time. A strategy was employed to get the operators and management together to reveal the data, present what the real departmental needs were, identify potential barriers and concerns, and fix what was needed to permit improvement. The results demonstrated a significant reduction in the amount of batch reworking, and increased production levels were achieved. This accomplishment was gained by all the people working on the process using problem-solving tools and statistical data analysis. The results have continued to improve without performance standards and slogans. These production people are trying to be better than they ever have been. Quality is up, cost is down, there's pride in the work, and everyone's having fun.

Point 11a. Eliminate work standards (quotas) on the factory floor. Substitute leadership. 11b. Eliminate management by objective. Eliminate management by numbers, numerical goals. Substitute leadership. Point 11 relates very closely to Point 10. Focusing on outcomes instead of the upstream process limits everyone's ability to meet customer demands at a satisfactory price. Work standards limit the potential for improvement and present a barrier to those working on the process. Work standards that encourage pay for good parts do not reflect the difference between special causes and common causes of variation. If you reinforce the work standard, that's what you'll get, not top-quality goods.

What is the job? With work standards the job is to produce a certain number of products per time period, not to meet the customer needs or continually search for new and better ways to do so. There may be forces in the process, such as raw material and measurement control, that act as inhibitors to work-standard achievement. People work in the system, doing the best they can. What is missing is management's help to improve raw material and measurement variability.

The following example seems appropriate. A particular piece of equipment was scheduled to produce a specified amount of pounds per day. However, the equipment frequently became clogged, which required a machine change-out. When these failures occurred, the people worked hard on the change-out to get the machine back into service. The focus, though, was on keeping the people working harder to maintain the quota. By using statistical thinking, a participative approach, problem-solving tools, and managing consequences, the entire process was put under scrutiny. The data were analyzed and shared with the people, and problems in methods of change-over were identified and corrected. Materials used in the equipment were studied and improved (made more fit for the intended use). People were recognized for doing things correctly. The results were beyond what was believed possible. Quality was up and costs were down—there was pride in workmanship and team spirit and there were happier customers.

Point 12a. Remove barriers that rob the hourly worker of his right to pride of workmanship. The responsibility of supervisors must be changed from sheer numbers to quality. 12b. Remove barriers that rob people in management and in engineering of their right to pride of workmanship. This means, inter alia, abolishment of the annual or merit rating and of management by objective. Examining each management system to determine if it supports or inhibits the company in meeting customers' needs is time well spent. Every management system is a process, eliciting responses, and supporting or limiting the people in the process. The new economic era demands that management understand concepts and sources of variability and make changes in the management processes to meet customers' targets consistently over time. Quality management gurus have stated that management's job is to get everyone's help who works in the system to work on the system. Changes must be made so that work being done today will not have to be done again tomorrow. Such management processes requiring scrutiny include cost-accounting systems, procurement systems, production and project reporting systems, innovation and marketing systems, human resource development and improvement systems, and resource-allocation systems.

The previously noted systems have both internal and external customers. They all require financial resources to operate, and each impacts the bottomline. If management really expects to meet customer needs at a price the customer is willing to pay, they must change these systems.

As an example, a marketing system for forecasting price and volume was studied. In the study, an industrial engineer worked with a management team trying to identify and correct problems with forecast inaccuracy. This study revealed that the marketing expense plan was managed independently. As a result, the resources needed to achieve forecast projections were limited so that the expense plan could be met. Thus, as long as the expense plan and forecast were not tied together, management could not expend the resources needed to meet the forecast. This had a great impact on production expansion and business planning. Studying these two processes simultaneously helped in removing a barrier to achieving more accurate forecasts. The results, although early in the game, reveal more consistent attainment of price and volume forecasts. This, in turn, has led to shipping schedules being met, a more deterministic business, more accurate business planning, and more faith in expansion plans.

Point 13. Institute a vigorous program of education and self-improvement. As Deming's philosophy is instituted and organizations strive for continual improvement, process changes to all facets of the organization will occur. This reinvestment in an organization can only occur if it reinvests in its people: top management, middle management, first-line management, support and staff people, and hourly workers. Management must make it perfectly clear they will reinvest in the people whenever necessary.

Each employee has to learn about the quality principles, concepts, and practices. In addition, the employee must understand existing physical and management processes and how they help or hinder customers' needs being met. Management's understanding of the human resource is the key to this success. Once this understanding is reached, a plan must be developed to invest in the people, including an assessment as to how the investment will be realized. Effectively using human resources is important for success, for making improvements, and for continuing to meet customers' needs. People must identify and eliminate those factors that rob people of their right to do a good job.

Point 14. Put everybody in the company to work to accomplish the transformation. The transformation is everybody's job. Deming's recommendation is very definite concerning organization transformation for meeting the challenge set forth in this new economic era. The transformation kit includes assigning or hiring an experienced, enthusiastic, master level statistician who is a good team player, obtaining a consultant's services to work with the statistician, and creating a common vision and customized strategy for guidance.

Every manager must be involved in leading and teaching the new management process. The responsibility cannot be delegated. Everyone must be willing to learn new ways of statistical thinking, new ways of securing and using employee involvement, and new ways of managing consequences. Those in command must be actively involved in executing this strategy, otherwise people will look to the consultant, statistician, or other staff resources. The new way must be modeled or practiced by every manager the next day and every day thereafter or it will be lost.

This model customized for the plant's environment must be executed by industrial engineers, statisticians, and other quality specialists. Senior executives must be involved in developing a mission and future vision and deciding on key areas for tracking business success. They must lead the way for their subordinates, teach them, help them get started, link-up on important areas of management concern, and work together more than ever before. These senior executives must envision all employees being involved, using the same language (statistical), solving problems, removing barriers, providing systematically earned recognition, and producing products that meet or exceed customer needs.

CONCLUSION

As stated previously, applying one or more of Deming's principles will lead to improved quality as perceived by the customer. Systematically applying these principles will ensure corporate survival. Clearly, there is an opportunity for all industrial engineers to improve contribution and value. Corporate survival depends on total quality leadership, focusing on the internal, external, and potential new customers. The industrial engineering profession measures its success by its impact on the business individually served. Technologies exist to survive, and industrial engineers are well suited for meeting the challenge. Get to work!

References

Daniels, A. C. and T. A. Rosen. 1983. *Performance Management: Improving Quality and Productivity Through Positive Reinforcement.* Performance Management Publications, Inc. Tucker, Georgia.

Deming, W. Edwards. 1986. *Out of the Crisis.* Center for Advanced Engi-neering Study, Massachusetts Institute of Technology. Cambridge, Massachusetts.

Juran, J. M. 1979. *Quality Control Handbook.* McGraw-Hill Book Company. New York.

Peters, T. J. and R. H. Waterman, Jr. 1982. *In Search of Excellence.* Harper and Row Publishers. New York.

Tribus, Myron. 1984. Creating the quality service company. Center for Advanced Engineering Study. Massachusetts Institute of Technology. Cambridge, Massachusetts.

Part IV. A Vision of the Future

Chapter Eleven
The Quality Imperative in the New Economic Era

Myron Tribus
Director
Center for Advanced Engineering Study
Massachusetts Institute of Technology
Yoshikazu Tsuda
Professor of Statistics
Rikkyo University
Tokyo

THE QUALITY IMPERATIVE

In the last thirty-five years a new principle regarding competitive advantage has emerged and is having a profound impact on international trade. This principle runs counter to generally accepted concepts in economics. Most economists seem unaware of its validity, origin, theoretical basis or its evident impact on international trade.

The new principle as yet has no name. Since so many concepts of economics are reminiscent of thermo-dynamics, I suggest it be called the "Zeroeth Law of Quality Management".

The zeroeth law of quality management: the producer of highest quality is likely to be the producer of lowest cost

Appreciation of this principle requires that the word "quality" be properly understood. As Pirsig has demonstrated in his delightful little book, *Zen and the Art of Motorcycle Maintenance* (Morrow Publisher 1974), the word "quality" cannot be defined by reference to more primitive ideas. It is itself a primitive idea. To explain it requires that examples be given of what is and is not meant when the word "quality" is used.

To begin, it is possible to say what quality is not. For example: *Quality is not a synonym for features.*

The features of a product or service are those attributes which are designed into the product to meet special desires of the customer. Features serve to define the market niche for which the product was made.

In homebuilding, for example, contractors will often say they are building a house of "high quality" when, in fact, they mean they are adding features to the house. A home builder once proudly explained that his houses were of high quality because they boasted air cleaners, music in every room, double glazed windows and similar expensive features. However, when I visited the owners of houses he had built I found that they complained of leaks in the roof, floors that were not quite level, and seepage of water in the basement. The builder used the word "quality" in a way that was deceiving to the uninitiated. People seem to think that if you order many features, you will automatically obtain a quality product.

An air conditioner added to an automobile provides a feature. If the air conditioner performs quietly, consistently and flawlessly, it may be said to provide quality.

Another example of a feature is color. When Henry Ford introduced his automobiles he used to say that the customer could have any color as long as it was black. The development of different colors for automobiles introduced a feature. When the paints were improved so that they did not fade in sunlight and when the finishes became smooth and mirrorlike, they added quality.

Quality is associated, therefore, with integrity in delivering what a customer has a legitimate right to expect in view of what was promised at the time of the agreement to purchase. Even this definition is inadequate.

Some people try to define quality by saying: *Quality is "conformance to specifications"*. [A poor definition.] What I believe they mean to say is: *Non-quality is non-conformance to specifications.*

As every student of logic knows, the definition of a *negation* does not lead to the unambiguous definition of an *assertion.*

Specifications cannot, as Pirsig has demonstrated, define quality because quality itself is a primitive idea. Specifications provide *legal definitions of acceptability.* In the marketplace to strive to be merely acceptable is to opt for last place. Quality producers know they must be constantly striving for improvement in quality.

To define quality as conformance to specifications is to confuse the concept of quality with acceptability.

TECHNOLOGY AND QUALITY
There is a difference between an art or craft on the one hand and a technology on the other. An art or craft is a personal skill which takes time and patience to master. It is associated with an individual person. Not everyone can develop the necessary skill, and it is found that some are much better than others. To obtain the skill, you have to hire the person.

A technology is a *technique* which has a scientific and logical foundation to it. In a technology, the essential skills are provided by a machine. Technology transfer consists of teaching people to understand the underlying logic or science and how to use the machine. A "high technology" is one in which the reduction to practical use requires advanced education.

We say a technology has been "mastered" when it is reduced to a "low technology" status, i.e., almost anyone can learn to use it.

An art or craft is applied in a "cookbook" style, in well described circumstances and without the need for understanding by the user. A technology, on the other hand, is expected to be applicable to many circumstances. It requires the understanding of the underlying theory and intelligence on the part of the practitioner.

Technologies do not only pertain to hardware manufacture. We can recognize at least five categories of technologies.

1. Technologies of products (to provide improved features)
[Aircraft., washing machines, computers . . .]
2. Technologies of processes (to provide qualities)
[Cutting, imaging, heating, fastening . . .]
3. Technologies of manufacture (to improve qualities)
[Robotics, automation, testing . . .]
4. Technologies of engineering (features and quality)
[CAD, CAM, optimization . . .]
5. Technologies of management (to improve quality)
[CPM, PERT, CWQC . . .]

Technologies of products enable new features, but they also make possible higher quality. The introduction of a truly "new" product usually involves a new feature. Steady improvement usually involves improved quality.

New technologies of manufacture often make it possible to supply a quality that was not attainable before. New methods of spray painting, for example, make it possible to produce new surfaces. New methods of microchip manufacture make possible more features in a small volume.

When an attempt is made to improve the quality of the output of a system by purchasing advanced machinery, we speak of it as a "technology fix". When U.S. manufacturers first became aware that Japanese goods were being produced with higher quality and productivity they thought it was a combination of cultural and technical advantages. When they went to Japan to investigate they were surprised to find that the equipment was the same as at home. Many of them concluded that the differences were "cultural". They did not find the "technology fixes" they had expected.

Garvin in an article entitled, "Quality On The Line" (*Harvard Business Review* 1983) has compared the propensity to make errors and produce flaws in comparable companies in the U.S. and Japan. He found that in similarly equipped factories that the error rate in the U.S. was 500 to 1000 times higher. Other observers have commented that in the U.S. flaws are measured in parts per thousands; in Japan they are measured in parts per million.

In 1979 Makoto Takamiya made a study of four factories in Great Britain (Discussion Series Int'l Institute of Management). Two were owned by Japanese, one by Americans and one by the British. In each case the ranks of management (except for the very top posts) were British. Table 1 shows the characteristics of the companies studied.

A Vision of the Future

Table 1
Cases Studied

	# OF EMPLOYEES	PRODUCT	METHOD OF MANUFACTURE
Japanese 1	700	color tv	large batch/mass production
Japanese 2	300	color tv	large batch/mass production
American	700	color tv	large batch/mass production
British	2000	color tv	large batch/mass production

Some "measures of success" are tabulated in the next table.

Table 2
Measures of Success

	J1	J2	A	B
Labor productivity (1)	0.83	1.07	0.71	0.56
Quality (2)	4-5%	10%	14-15%	85%
Employee satisfaction (3)	15.6	12.7	13.2	11.32
Labor turnover	30%	25-30%	30%	30%
Absenteeism	4%	5%	8%	9%
Days lost to strikes (4)	0	0	0	20

Notes:
(1) Sets per day per labor day
(2) Reject rates of printed circuit boards assembled
(3) Employee surveys
(4) Previous two years

Dr. Takamiya remarks that the popular image is that Japanese managers create a "family" in which the workers find greater happiness and that this contributes to the greater productivity and quality. In the surveys, however, the morale of American managed workers was negligibly different from the Japanese managed workers. Deliberate attempts to make workers unhappy undoubtedly can lead to lower productivity. However, efforts to increase worker happiness per se does not automatically increase the productivity of the system. Table 3 reviews the incentives used in the four companies.

"Technology fixes" did not explain the differences in productivity either, as indicated in table 4.

Dr. Takamiya attributes the differences in productivity and quality to *management* and identifies these areas of difference.

Product selection. The British had 60 types of products, the Japanese had 4 or 8, and the Americans 6.

Attention to "minor" details. The Japanese were meticulous with respect to details. American and British managers were lax.

Table 3
Measures of Worker Satisfaction and Motivators for Satisfaction

	J1	J2	A	B
Worker satisfaction (1)	7.82	9.61	12.40	12.18
Sick pay	0	0	yrs. of service: weeks 1-2: 4 2-3: 6 3-4: 8 4-5:10 5-6:18 6-7:28 7-8:28+6 8-9:28+12 9-10:28+18 10- :28+24	2-5: 8 5-10:16 10-15:24 15-20:32 20- :40
Holidays for hourly workers	after 1 yr. (fixed period)	after 1 yr. (fixed period)	20 days	<1 yr. 20 days 1-3 yr :21 4 :21 5 :22 6 :23 7 :24 8 :25 12 : 25+ one weeks basic pay

Notes:
(1) Employee survey

Table 4
Use of Technology

	J1	J2	A	B
Auto insertion equipment for circuit boards	Y	Y	Y	Y
Circuit board testing	manual	manual	manual	automatic
Transport of goods	human	conveyor	conveyor	conveyor
Inspection	eyes	eyes	eyes	machine

Work practices/supervision. Japanese worker assignments were flexible. If necessary secretaries would fill in for workers as occurred in one emergency. By contrast, the jobs of workers in American run plants were governed by "job descriptions". In the British run plants, jobs were determined by agreements with the union.

Discipline. The Japanese insisted on strict discipline. Both the Americans and British were much more relaxed.

Interdepartmental coordination. The Japanese plants had strong interdepartmental coordination, enhanced by policies of frequent rotation of managers among departments. Interdepartmental relations were more formal in American run plants. In the British run plant some departments considered the other to be the "enemy".

It was observed that in Japanese run plants the employees were responsible to clean up their own areas. They were involved in job improvement, were constantly monitoring their own quality and involved in discussions with supervisors about improving the workstation. There was a turnover of 28% in new hires in the Japanese plants which Takamiya attributes to encountering the strict discipline.

The Takamiya study makes it quite clear that the *adoption of managerial technologies* makes the difference. Except for a few top spots, the management of all four companies was British. These British managers had adopted the technologies of their bosses.

- The advantages of the Japanese companies in Great Britain is not cultural.
- Their advantage lies in the use of improved managerial technologies by upper, lower and middle managers.
- These technologies are transferrable to managers who choose to learn them.

THE COST QUALITY CONNECTION

The foregoing discussion has laid the basis to explain why the quality principle works and why it is of such importance in economics.

Suppose two producers design their products for the same market niche. They present the customer with the same choice of *features*. The customer will decide which one to purchase based upon perceptions of quality and cost. Suppose the product is a fire extinguisher for use in the home. Let both have the same capacity, as testified to by Underwriters Laboratory on the label. Suppose one product has a smooth finish and the handle is free of burrs and scratches while the other shows an "orange peel" effect in the painting and the handle is not so smooth. The customer will certainly prefer the product that exhibits superior craftsmanship. This is a subtle indicator of the aspects of the product that cannot be seen.

If we were to go to the factory of the first producer we would be likely to find that the spray system is much cleaner than the competitor's. The person doing the spraying will have been trained to mix the ingredients of the paint so as to get the viscosity correct, to keep the spray gun the proper distance from the product, to keep the system clean and to inspect it periodically. It will probably be found that the competitor has not trained the operator properly and does not know what viscosity or spray pressure is best. The equipment is likely to be down for unexpected repairs. In short, the quality of the product reflects how well the manufacturing process is managed.

Can anyone who examines the data of Garvin or of Takamiya doubt that when 80% of the product requires rework before it is considered *acceptable*, the resulting output will be of higher cost and lower quality? This fact of life is the basis for the quality principle.

Managements that learn to manage for quality have an unbeatable advantage of those who do not.

There are some who consider quality to be an end, not a means. They are apt to say: "Quality is free". This is a misguided statement. *Quality management is a means to low cost and high productivity. To say that "quality is free" is to confuse means with ends.* This statement goes against the first principle of quality management which states: *Quality is never your problem. Quality improvement is the answer to your problem.* Quality improvement is not an end. It is a means.

The "leverage" of quality has been pointed out by Feigenbaum in his book, *Total Quality Control* (McGraw Hill 1983), who referred to the "hidden factory", i.e., the factory within a factory where employees were busy fixing the mistakes of the main factory. This hidden factory is busy taking care of

Table 5
The Economic Leverage of Quality

	1983 ANNUAL REPORT	1983 ANNUAL REPORT (RESTATED)	50% QUALITY IMPROVEMENT	7% NET NEW BUSINESS
INCOME	12,581	12,581	12,581	18,309
EXPENSES				
Cost of products sold	8,020	6,416	6,416	9,281
Cost of poor quality		1,604*	802	2,320
Other expenses	3,011	3,011	3,011	4,356
	11,031	11,031	10,229	15,957
Gross Earnings	1,550	1,550	2,325	2,325
Income tax	684	866	1,085	1,085
Net Earnings	866	866	1,267	1,267

* Estimate of a typical company
  Tax rate = 50%

rework, scrap, inspection, management meetings to see what to do to fix whatever went wrong, supporting sales people placating customers, purchasing agents who are trying to get emergency supplies because someone ordered or delivered the wrong thing, etc., etc. Feigenbaum suggests that this factory amounts to 20% of the company personnel and accounts for about 20% of the cost of manufacture.

The competitive consequences of the decreasing cost quality curve are very great. Heretofore it was common experience that to raise the quality of one product over another, it was necessary to increase the cost. It was possible, therefore, for economists to speak of the market as "segmented," considering in one segment products of higher quality and higher cost and in another, products of lower quality and lower cost. It is still possible to segment the market according to *features* but it is now necessary to take into account the *destabilizing* effect of the inverse cost-quality function.

The new breed of managers has learned that it is possible to lower costs by increasing quality. This means that the higher quality producer has a double advantage. The higher quality product can command a higher price, yet because it has a lower cost to the producer, it gives the enterprise a much larger profit margin than its competitors. It can use this margin in several ways, including the option to finance growth internally as it moves into new product lines. It can also keep the competitors on the verge of bankruptcy as it slowly lowers its prices and gains market share.

This sort of thing has now happened in automobiles. By applying the new management techniques Japanese automobile manufacturers have attained a significant cost advantage over American producers. The following table is taken from a talk by James Harbour and indicates the situation a few years ago, just before significant managerial changes were made at some of the American automobile companies.

When studying Table 6 it should be kept in mind that what the table does not show is that the managerial practices which lead to these cost advantages *also produce quality advantages. The Japanese cars not only cost less, they are of higher quality. They cost less because they are of higher quality.*

Table 7 explores some of the consequences of operating at higher quality by providing a comparison between two similar plants in the U.S. and Japan, each concerned with stamping out metal parts and assembling them.

When management puts its efforts into making the work go more smoothly, there is less scrap, less rework, less milling around, less wasted human effort, greater employee morale, greater creativity on the part of all workers, greater job satisfaction, greater productivity, greater customer satisfaction ... In short higher quality, productivity and stronger competitive position.

All over the globe managements are learning the quality principles. At the Center for Advanced Engineering Study we see a steady stream of foreign visitors, eager to learn about the new style of management. Organizations of enthusiasts are forming in Europe, in Latin America, in Asia. Through an organization called the Association of Overseas Technical Scholars, Japan is exporting this kind of management to 17 Asian nations.

Table 6
Cost Advantage ($/Car) of Japanese Auto Manufacturers Over U.S.

CATEGORY	COST ADVANTAGE $/CAR
Labor (60 vs 120 hours)	550
Advanced Technology (greater use of robots)	73
Quality Control (more inspectors, repair men)	329
Excess Inventory (work in process)	550
Materials Handling (extra manpower)	41
Better Use of Labor (fewer job classifications)	478
Absenteeism	81
Assembly Line Relief System (shut down vs "tag")	89
Cost of Union Representative	12
Total Cost Advantage	2203
Shipping, Handling, Import Duties	-485
Total Cost Advantage	1718

Table 7
Some of the Cost Advantages of an Operation Managed For Quality

	JAPANESE	AMERICAN
Daily Production	1000	1200
Size (millions of square feet)	1.5	3.0
Direct Labor (people)	1112	2600
Indirect hourly (people)	625	2125
Indirect salaried (people)	175	525

Their motivation to do this is quite simple. Japan expects to move offshore to other countries those industries in which the competition has learned about quality and thereby reduced their competitive advantage. They expect to retain in Japan those industries which have a very high value added per pound of product. They intend to stay ahead of the race by concentrating on high technologies. This is a very intelligent strategy for a country which has only people as a resource and must import essentially all the materials from which it makes goods for export.

QUALITY — THE ECONOMIST'S BLIND SPOT
The result of the learning now going on will be that soon the quality principle will be used all over the world. Because international trade has become essential to the survival of all nations (including the USA), no company will be immune to the increased competition.

Any economist who intends to provide useful forecasts and guidance to a government needs to take into account this new development.

Economists make their predictions and render their judgements by studying *economic indicators*. These indicators are taken from the numbers which represent *quantitative factors* in the economy. The objective of economic theory is to provide insight and understanding of what is happening and what is likely to happen in the economy, based on data about the economy itself and whatever other information is available. Until the quality indicators are included in the theory, predictions will be inaccurate, even misleading. Existing economic theory simply would not have forecast the Japanese ability to import essentially all the raw materials it needed and then to compete with a fully developed automobile industry established in a country which needed very little importation, and to beat the industry on its own ground. It has happened with one industry after another in one country after another. The quality principles cannot be ignored any longer if economic predictions are to be of any use.

The development of economics has been greatly influenced by the concepts in thermodynamics. For example the concept of "equilibrium" is discussed in terms which are quite similar to those used in discussing chemical equilibrium. Nicholas Georgescue-Roegen has gone so far as to suggest, in his book *The Entropy Law and the Economic Process*, that a true theory of economics should be based upon the laws of thermo-dynamics, especially the second law (Harvard University Press 1971). Whether this is true or not is arguable, but his book does make clear how similar many of the arguments of thermodynamics and economics are.

Economists make good use of the conservation principles of mass and energy. For example, the basic "input-output tables" for which Leontief received a Nobel Prize is a matrix which accounts for the fact that in all manufacturing processes there is a conservation principle which can be used to see how the change in activity of various industries changes the demands upon the sources of these materials and one another. The Nobel Prize Committee recognized the importance of this contribution to our understanding. The First Law (or more generally, conservation principles) provides limits to performance. You cannot make something for nothing; you cannot create energy, you can only obtain it from a source. An input-output table provides information from which it is possible to see how well the transformations occurred.

In real processes of manufacture, what is actually accomplished is only a small fraction of what could be done. The theoretical limits of productive capacity are very much greater than what actually occurs. If the productive efficiencies of the future are assumed to be the same as those of the past, predictions about the future will make about as much sense as driving a car down the highway guiding it by looking at the white line in the rear view mirror.

For an economic forecast to be relevant, it should take into account two factors which have a strong effect on the economy. The first is *technology* and the second is *quality*.

Competitiveness derives not only from natural advantages i.e., proximity to natural resources, but also from two thrusts:

- Invention and innovation, making new and better things.
- Quality and productivity, making things in better ways.

New technologies of products, processes and manufacture strongly influence invention and innovation in products. The Congress seems to see a close connection between the rate of innovation and the support for research leading to new and improved technologies on the one hand and international competitiveness on the other. The USA has had a long standing tradition of trying to support "Yankee ingenuity".

New technologies of management strongly influence quality and productivity. Because so many people look upon management as a privilege, it is difficult for many of them to accept that if they use an improved technology, they are not admitting that they were inadequate in using old technology. At any rate, for a given level of technological advance, it should be obvious that the way the system is managed determines the quality of the output.

Until very recently, economic theories did not take into account technology. Some economists are revising their approaches to make technology an explicit factor in their equations. To take into account technology will not be enough. Developments in quality management now have to be included as well.

As a thermodynamicist I cannot help but be struck by the similarity between the role of *quality* in economic competition and the role of *entropy* in thermodynamics. Entropy is a useful measure of the potential to improve things. Gain in entropy means loss of thermodynamic quality. Gain in entropy also means loss of information. As we shall see, loss of information during manufacture also means loss of quality. The formal aspects of this intriguing analogy remain to be explored. They are potentially of great interest. These considerations are beyond the scope of this paper.

The new era of economic competition is dominated by the quality imperative: *Learn to reduce costs through increasing quality or sell out to those who already know how.*

THE MANAGERIAL CHALLENGE

Some people refer to the impact of quality as the "Second wave of the industrial revolution". Before the industrial revolution goods and services were produced by skilled artisans working by themselves or with apprentices. In the case of larger projects, the work was done by a hierarchy of managers and workers, under the leadership of a skilled person. The individual tasks were relatively simple and unchanging. Certain crafts could be identified: bricklaying, stonecutting, concrete mixing and pouring for example. The industrial revolution saw the skill of the individual craftsman transferred to the factory, where the work could be better organized. In United States in the 1930's the industrial revolution had progressed to the point where two kinds of unions were formed. The AFL was organized along traditional lines, according to crafts. The CIO was organized according to industry. As we shall see later, the full impact of the process of industrialization occurs when the skills of the individual person are transferred to the *organization*, not just to the place of work.

Wherever strong international competition is occurring the number of job classifications is being severely reduced. Workers are expected to move from one task to another, filling in and helping one another as required. Whereas before people associated themselves with specialist jobs, now the emphasis is shifting. This shift requires that the jobs be described in terms of *standard procedures*. In a modern factory the practice is to post at the place of work a written description of how to do the task, presented in such a fashion that a person can slip into position and begin to work with a minimum of training. Complete interchangeability of workers is not possible, of course, but that is the ideal—the complete transference of the skill to the organization.

In such a setting it is important that the human not be diminished. Managerial approaches which respect humanity are the hallmark of the successful manager.

Until recent times, top management and schools of management have paid very little attention to technologies aimed at increasing quality. This is certainly not because the basic information was unknown. The theory was laid down by Walter Shewhart in *Economic Control of Quality in Manufactured Product* (Van Nostrand 1931). Statisticians such as W. Edwards Deming and Joseph Juran have carried the message around the globe ever since. Dr. Deming's book, *Quality, Productivity and Competitive Position,* is devoted to the changes managements must make to apply the technologies for quality (MIT 1982). But their teachings were ignored everywhere (except Japan) for a half century. It is easy to look back and understand why it happened. In the era just after WWII the emphasis was on quantity of product, not quality. Now people have to learn new ways.

Most people seem to learn a new outlook on life only when they must. Most of us tend to lead unexamined lives, basing our everyday actions on custom and usage. When our beliefs are challenged, most of us react with anxiety and sometimes even anger.

Philosophies of management go back a long way in history. Stucart and Eastlick have written a brief history of the development of managerial concepts, starting from the building of the pyramids (Libraries Unlimited 1981).

The Old Testament described an hierarchical organization in "Exodus", indicating that Moses "chose able men out of all Israel and made them heads over the people, rulers of thousands, rulers of hundreds, rulers of fifties and rulers of tens".

The hierarchical form of organization was used by the Romans in conquering the entire known world 20 centuries ago. When reading this historic review, two things seemed to leap off the pages:

1. Historically, management has been a privilege
2. Labor has been a commodity.

A cartoon from the MANS (Management, A New Style) Organization in Holland shows this is not a unique deduction:

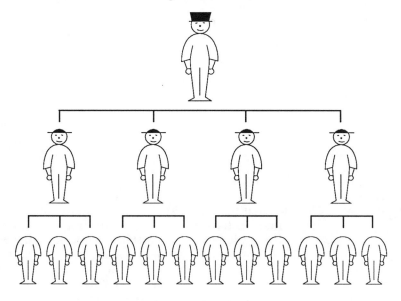

Figure 1. A Hierarchical Organization

At the start of the 20th century, this image of what it means to be a manager was given a certain intellectual status by the work of Frederick Winslow Taylor, who became known as the "Father of Scientific Management". Here are two quotations from his work:

> "Hardly a competent workman can be found who does not devote a considerable amount of time to studying just how slowly he can work and still convince his employer that he is going at a good pace".

> "Under our system a worker is told just what he is to do and how he is to do it. Any improvement he makes upon the orders given him is fatal to his success" (Harper and Brother 1911).

In later years Taylor was to lament what had been done in the name of "scientific management". He preferred to dwell upon the intellectual and scientific contributions (i.e., time and motion study, work measurement, methods engineering) which he had pioneered and did not wish to be associated with the sociological effects.

Juran has called the hierarchical or "tree structure" the "Pyramid of Power" and has pointed out that the nature of the conversation changes as one ascends the pyramid. At the bottom people talk about things. Near the top they talk about dollars. The people in between are translators, who convert things (programs) into dollars (budgets) and vice-versa. This is an oversimplification, of course. Unfortunately, it is not that over-simplified. The translation of things into dollars is the way that top managements connect what happens in the factory with what happens in the macro-economic world outside the factory (see Figure 2).

Ralph Landau, a distinguished chemical engineer, has written a strong attack on the state of economic theory. One of his main points is that while what happens at the level of the factory determines what happens at the level of the national economy, there is no theory that ties the micro- and macro-economics together. He points to physics in which it is understood that the behaviour of very small particles is described by quantum mechanics while the behaviour of systems of particles is described by classical physics. The connection between these two views is made through statistical thermodynamics.

In economics, unlike physics, the two theories are not connected. There is no micro-theory and connecting analog to statistical thermo-dynamics. It appears to me that the difficulty cannot be laid to the door of the economists. They, after all, have to work with the figures that are available to them. The difficulty lies in the understanding of the managers themselves as to what it is that influences the performance of the organizations they are supposed to lead. Most managers think in terms of a tree structure. This image of the organization needs to be changed if a manager is to learn the technologies associated with quality improvement.

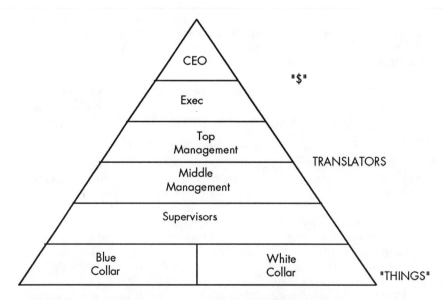

Figure 2. Languages In The Pyramid Of Power

CHANGING THE MANAGER'S VIEW OF MANAGING

To begin, we define the manager's job as follows: *The people work in a system. The job of the manager is to work on the system to improve it with their help.*
Managers should not visualize themselves as sitting atop a portion of the pyramid. Rather they should think of themselves as responsible to improve the quality of the performance of systems of people, machines, processes and procedures.

This change in perspective cannot be mandated. It cannot be "managed". It must be led. If top management thinks of itself primarily as engaged in a struggle for power to be at the top, so will all the the managers beneath them. In my opinion this has been an important factor in the decline of the British ability to compete in world markets. The top positions in British industry have been held for so long by people of privilege that they have lost touch with what life is like near the bottom of the pyramid.

Something akin to the British experience has been brewing in the USA. For about 35 years American schools of Business Administration and Management have been turning out MBA's for whom entering the pyramid of power has been a guiding ambition. They have been "groomed" for "top management". The prestige schools make no bones about it. Their ambition is to produce the leaders of tomorrow's industry. Unfortunately, their training is aimed mostly at conquering the pyramid. It is certainly not aimed at starting at the bottom and straightening out how the work gets done. When the top of the pyramid is concentrated on the flow of dollars, there is a tendency in good times to take the maxim: *If it ain't broke, don't fix it*. In stringent times, there is no time to fix things. This maxim needs to be replaced with a different one: Continuous improvement is a way of life.

To take a place in a quality enterprise, the managers need to develop new competencies and a different "image" of the enterprise. They need to learn a new philosophy of management and new managerial technologies.

The new philosophy of management requires a reconsideration of the purposes of the organization, the rights, responsibilities and obligations of all the people in it, and, especially, how managers and workers should relate to one another if the organization is to be competitive in world markets where quality is a driving force.

The new managerial technologies enable managers to improve systems, to increase quality (thereby cutting cost) and to smooth the flow of innovative products and the introduction of new processes. Through these technologies managers can better serve the customers and gain market share while becoming more profitable.

THE OBJECTIVES OF THE NEW STYLE OF MANAGEMENT

A manager must satisfy a wide variety of "stakeholders". The enterprise looks different to different people, depending upon whether they are in the organization or outside it. Figure 3 gives a picture of how different people with different interests in the organization perceive it.

Competitors, customers, news reporters, financial analysts and shareholders all are on the "outside". What they see is just the tip of the iceberg. If it is a quality company they see remarkable new products, good value for

A Vision of the Future

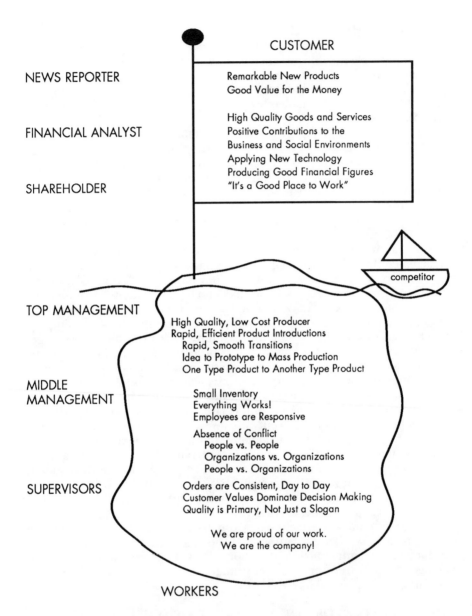

Figure 3. The Many Views of the Corporation

the money, high quality goods and services. When they deal with salespeople they are impressed at their understanding. Customers feel that those who serve them really care about doing a good job.

Shareholders see that the company is applying new technology and at the same time producing good financial figures. The company is making good contributions to the business and social environment. The company is seen by reporters as a "good citizen" in the community. They see the company growing in influence. The management of the company is invited to serve on committees of the community, and people are pleased when they do.

The financial analysts report that the company is doing well, is a good investment, safe and with good growth potential. Competitors see it as a formidable force in the marketplace.

Top management sees that the company is a high quality, low cost producr. The company has a good track record in the introduction of new products, bringing them out on time, within budget and free of "call backs". The products are obviously designed to meet market needs. When it is necessary to change from one kind of product to another, the changeover goes smoothly.

Middle management sees that the inventory is small and that there is a rapid turnover of inventory. Things work; the systems and procedures do what they are supposed to do. Departments do their work swiftly and efficiently. The equipment is well maintained and adequately modern. The employees are cooperative. There is a spirit of "can do" about the place.

Supervisors do their work in the absence of conflict. They are not caught in the middle, trying to explain why today's orders are in conflict with yesterday's orders. They generally know what to do without being told because the purposes of the company are clear. Customer values dominate decision making.

The workers see that quality is important. It is not just a slogan. The management is attentive to quality, and everyone is helped to do a better job. Management is visible everywhere and does not hide out in its upper offices. Supervision is helpful and employees are urged to find better ways to do their jobs.

The workers are rightfully proud of their work. They say, "We are the company". They trust the management to look out for them. They do not fear for their jobs. They believe their personal fortunes are bound up with the company fortunes. "We are all in the same boat".

Very few companies meet this image of the excellent company. In the next section we shall consider how to make an audit of your company to see where it stands.

HOW TO TELL WHERE YOUR COMPANY IS ON THE ROAD TO EXCELLENCE

When you know what to look for, you can learn a great deal about a company by just walking around and observing things. Since we are talking about a system of people, machines and procedures, it is important to observe all three.

When you attend meetings:

1. Do people present only the good news?
2. Do they report bad news only on request?
3. Do they present only raw data?
4. Are data analyzed statistically?
5. Are data presented in tabular form or as graphs and charts?

A Vision of the Future

How topics are discussed in meetings is a good indicator of where the enterprise is heading. In figure 4 we indicate some of the indicators as measured by how the management views data in meetings.
When you talk with the workers:

1. Do they spend time looking for ways to improve?
2. Do they take data on what they are doing and then discuss the significance of what they have learned?
3. Do they do their own housekeeping and are their work areas clean?
4. Do they regard the next person in line as a "customer" and try to understand the customer's needs? often?
5. Are safe practices the rule? obeyed?

Managerial attitudes influence the attitudes of the people reporting to them. Anyone walking through an office or factory can spot things which are not quite right. How the people react to you when you point them out provides considerable evidence as to where the company is headed. Figure 6 shows how management attitudes towards defects, errors, deficiencies and problems can be used to gauge the direction of the company.

An excellent company can be vulnerable. When you are at the top there is no place to go except down. Bad managerial practices can ruin an otherwise excellent company. Figure 7 shows some of the management actions, or rather inactions, which are the telltale signs of heading for bankruptcy. The most important of these actions are those in which the management does not want to see the problems.

These inactions are also likely to be revealed by management attitudes towards opportunities for improvement. The road to bankruptcy is paved with managerial incompetencies. Some of these are pictured in figure 8.

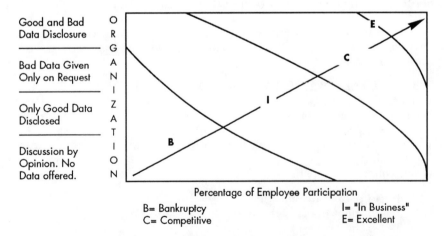

Figure 4. Managerial Attributes Leading to Excellence

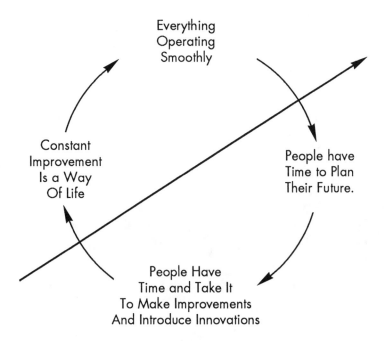

Figure 5. Management Action—The Road to Excellence

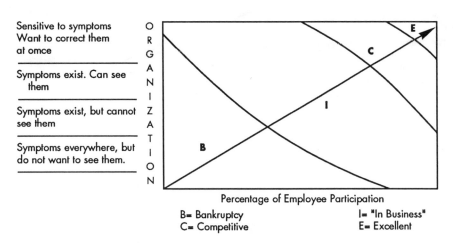

Figure 6. Managerial Attitudes Expressed in Meetings

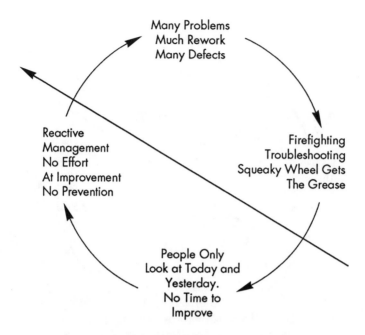

Figure 7. Management Inaction — The Road to Bankruptcy

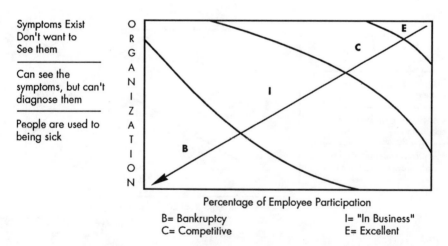

Figure 8. Managerial Practices Which Lead to Bankruptcy

If you walk through the enterprise, you can see if the management understands the pitfalls on the road to excellence. Some of the traps are pictured in figure 9.

It is easy to "walk the shop" and look for these kinds of evidence. A discussion with the workers will tell you if they are working to procedures. Talk with someone taking data and ask if they know what the data are used for? When did anyone ask? Do the numbers go anywhere besides the computer?
where their time goes, you can see if the important tasks of the enterprise are being taken care of. All managers have three basic responsibilities. These are:

1. To *maintain* the operation. That is, to tend to daily operations, keep things going.
2. To *improve* the systems for which they are held responsible.
3. To provide for the *future*, that is, to foster *invention*.

The degree of responsibility in all three functions varies with the level in the organization. Figure 10 shows a suggested split of responsibilities versus level in the organization.

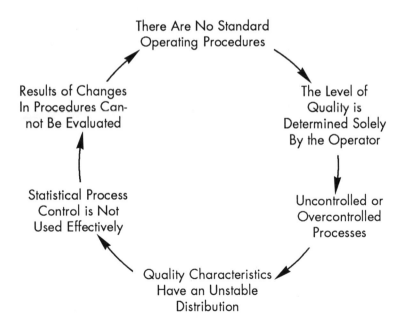

Figure 9. Potential Traps on the Road to Excellence

A Vision of the Future

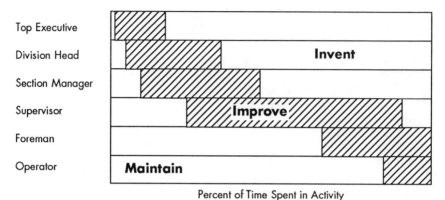

Figure 10. Management Divides Responsibility According to Organizational Levels

When the management does its job and the employees are all involved in improvement, a company is at its strongest. It is not possible to compensate entirely for one or the other. This balance is portrayed in figure 11.

THE TOOLS OF QUALITY MANAGEMENT

In the preceding sections we have discussed the obligations of management to press for continuous quality improvement. This pressure must be felt across the entire organization, not just on the manufacturing floor.

Since the manager's job has been defined as continuous improvement of systems, our review of the available tools begins with a discussion of different ways to approach the improvement of systems.

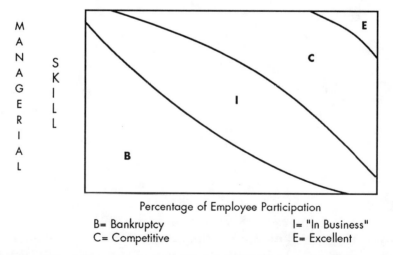

Figure 11. Managerial Skill and Employee Participation

To improve a system in an organized way requires several steps. These steps will seem intuitively obvious, but experience shows that most managers do not follow them. To improve a system you need to know the following:

1. What the purpose of the system is. What do people think the system is supposed to do? How is it judged? By whom?
2. How the system is supposed to work. Does it work the way it is intended? What goes wrong? How do the people cope with whatever goes wrong?
3. How can you define a quality output? Who decides? The people working IN the system know how to tell? Do their orders permit them to do a good job?
4. What potential exists to improve the system?

The First Tool: Fishbone (Ishikawa) Diagrams
If you want to know what excellence is, you have to go to whomever is the judge. In general this is your customer.

Many people believe they are far removed from the "customer", working deep within the organization. Quality conscious people realize that everyone in the organization is a "customer" of someone else. Once this fact is recognized, it becomes possible to identify the customers of the system.

I shall use an example from our Center for it illustrates two things. First, that the methods we are discussing can be applied in an office environment. Secondly, that the organization chart is a very poor guide to knowing how to improve quality.

In our Center the Headquarters staff is made up of about a half dozen people. There are three active programs in the Center, each program under a Program Director. When the headquarters staff met to discuss our job it quickly became evident that our customers were the people running the programs. We decided that a quality job from headquarters consisted in giving strong and effective support to our programs. We drew an Ishikawa or "fishbone" diagram (often called a "Cause and Effect Diagram") to illustrate how the various services we rendered came together to provide good program support. This diagram is reproduced as Figure 12. If you were to come to our Center you would find this diagram on a blackboard with different people's handwriting on it. Each member of the staff was encouraged to enter ideas on the diagram. These ideas were discussed in staff meeting. The Program Directors were invited to discuss the diagram. In this way we could identify what our jobs were and how our customers perceived what a good quality operation was.

Ishikawa, or "fishbone" diagrams are a powerful tool for managers. They provide a method to involve all the people in the improvement process and to see how the various factors come together to make up the total performance.

The Second Tool: Flow Charting
A fishbone diagram helps to relate the elements of a process, but it does not really show how the process works. To see how the people and organizations

A Vision of the Future

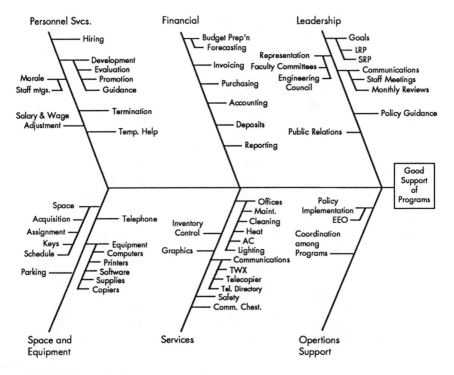

Figure 12.

interact with a process, a flow chart is helpful. One of the tasks we undertake in our office is to arrange for the annual meeting of our board of advisors. Arranging meetings, especially if they are infrequent, can be an awful chore. Our advisory board meets once per year and is drawn from a number of places around the company. Finding a suitable time in everyone's calendar is difficult, especially if the membership is voluntary. The responsibility is divided among several people in our office. Figure 13 shows a flow chart prepared by one of our secretaries, who did this after about four one hour sessions in the group. The flow chart was discussed among five of us and as a result, we were able to establish a firm procedure. This is a simple, almost trivial task, but when we worked together to understand it, most of the frustration we had felt around the office each year when it was time to call a meeting vanished. The flow chart provided the basis for a *standard procedure*.

Many people object to "standard procedures". They feel that such procedures stifle creativity. However, when these procedures are not imposed from above, but are worked out by the people themselves, as a way to do a job consistently and excellently, they do not appear as a constraint. Everyone understands that at future meetings the procedures are subject to review and improvement.

A flow chart could be developed to show the introduction of a product. This form of flow chart has one important advantage: It shows the

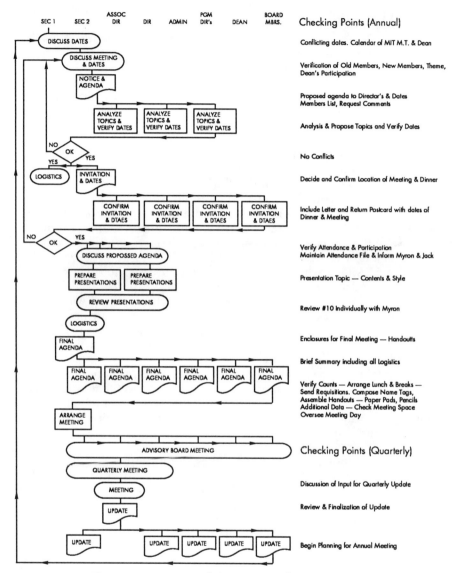

Figure 13. CAES Advisory Board Meeting

relationship between the people and the work to be done. When studying a flow chart a manager can ask the following questions:

1. Who are the "customers"? The flow chart identifies suppliers and customers by the horizontal transfer lines. Whenever a horizontal line appears it identifies a transaction involving a customer and a

supplier. The manager can ask if they have identified what constitutes a quality delivery.
2. What is the quality of the work? At every transfer point, if quality has been defined, it should be possible to take data on the quality of the goods or services delivered.
3. What are the barriers to good quality work? The people who work in the system can be asked to comment upon the flow chart. Does it actually show what they do? Can they show, by reference to the flow chart what their problems are?

Flow charts make the situation come alive. In an extremely thought provoking paper, F. Timothy Fuller has shown how a flow chart can illuminate a situation that no other way can do (Fuller 1985).

Figure 14 shows the flow chart for a simple assembly operation. It is a simple task, merely to take a kit of parts from the input station, assemble the three parts into a new part and place it in the output station. From the perspective of the manager, it is easy to do and appears as a simple task on the manager's flow chart.

Figure 15 shows what the task looks like in case the quality of the incoming kits is poor. If it is not certain that all the parts are there, the operator must first check if any are missing. Depending upon which ones are missing, a different assembly procedure is followed. Furthermore, a new procedure has to be followed by the supervisor, as indicated on the diagram. What was a simple task is now seen to be complex with many more possibilities for error introduced.

What makes a flow chart so useful is that once prepared it may be studied by all who participate in the process and improvements suggested. In presenting this flow chart, Fuller pointed out ways to improve it which, at first glance, would seem to be contrary to common sense. His paper makes for thoughtful reading.

A Third Tool: Nominal Group Techniques

In the preparation of fishbone or flow charts, it is helpful to involve the people who actually do the work. Until they become habituated to working in problem solving teams, it will be found that they are often unable to contribute. There is a skill in working together to solve problems and most people do not have it without some training. Most managers have not been

Figure 14. No Complexity

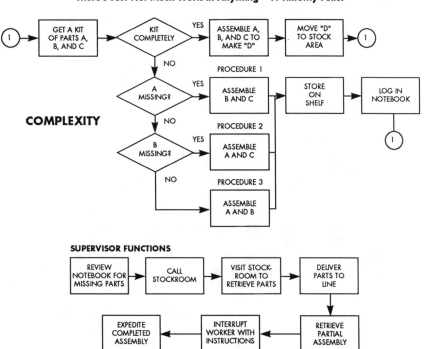

Figure 15. A process with errors. If an external quality problem is introduced into the assembly process, many extra steps are required.

taught to lead a problem solving team, and are uncomfortable and unskilled at it.

Nominal Group Technique (NGT), is a way to lead a group in problem solving. The steps in NGT are simple:

1. Silent Generation of Ideas.
    a. The group is asked to remain silent for a brief period of time and to write down on a piece of paper ideas concerning the question that has been posed by the leader. (The question could well be, "What is the most important question for us to consider?")
2. Round Robin Feedback.
    a. Each person is asked, one at a time, according to the seating arrangement, to present one idea from his or her previously prepared list. Each person is limited to one idea at a time. In this way, no one can dominate the conversation. Everyone gets a turn. When a person has run out of ideas, that person

says "I pass". The leader continues to poll the group until all say "I pass".
   b. As each person presents an idea, it is written on a board or flip chart. The ideas are presented as brief statements without elaboration. The person writing the statements is not to editorialize or "improve". It is important that each person's idea be written down as originally presented.
3. Group Clarification of Ideas.
   a. AFter the group has finished generating ideas, it will be found that many ideas are similar. Members of the group are asked by one another to clarify their ideas and the leader attempts to group ideas in similar categories, with the help of the members. During this process it will be decided by the group that certain ideas are so similar that they can be made into one. Such clustering should be done only with the consent of the persons who generated the ideas. The leader should be careful that no one is dominated by the group or has an idea lost.
4. Individual Voting on Ideas.
   a. If there are N ideas, each person is allocated a total of (1 + 0.5N) votes. That is, if 46 ideas were generated, each person will have 24 votes. Call the number of votes, V. Each person chooses the V most important ideas from the list and ranks them. For example, from the list of 46 ideas, each person selects 24 and ranks them in order of importance, putting a "24" opposite the most important idea and a "1" opposite the last.
   b. The leader tallies the votes. The tally is done by counting not only the scores, but also the number of votes. Someone reads from the slips of paper as the leader writes to the LEFT of the idea the NUMBER OF VOTES and to the right the NUMBER assigned. After all votes have been tallied, the totals are computed. Each idea will have been ranked in two ways: To the left will be a number showing how many people thought it was an important idea. To the right will be a number telling how strongly the voters feel.
5. Presentation in a Pareto Diagram.

A Fourth Tool: Pareto Diagrams
A Pareto Diagram is merely a bar chart in which the bars have been arranged in descending order, with the largest to the left. Pareto charts are useful to help a group set its priorities. Figure 16 show a pareto diagram for the time spent in different tasks by a group of office workers. On the left is the amount of time spent (measured in average annual wages) and on the right is the cumulative percentage of time. It can be seen that the total of time spent on typing, answering the telephone and retyping accounts for about 70% of the effort. Without a Pareto chart it is difficult for a group to agree

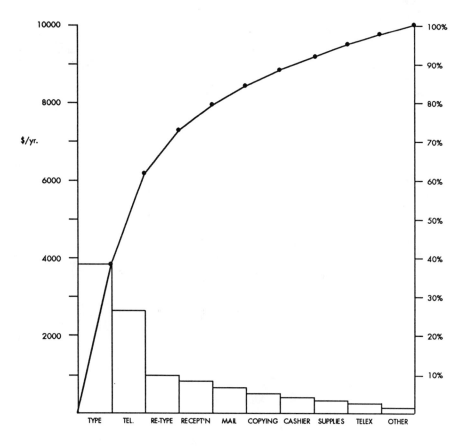

Figure 16. Activities of Clerical Personnel

on what to do next. Pareto Diagrams are an important aid to consensus building, based upon "fact based decision making".

A Fifth Tool: Run Charts
Run charts are merely plots of how a variable or attribute of interest varies with time. Successive observations are plotted on a graph. Figure 17 shows a run chart.
    The importance of run charts is that the eye is very good at noticing patterns. It takes much more experience to notice patterns when numbers are displayed in tabular form.

A Sixth Tool: Histograms
Figure 18 shows the histogram which resulted when an automatic machine tool was running freely. Figure 19 shows the distribution when the machine

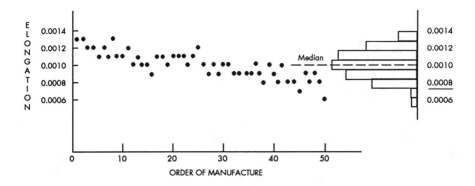

Figure 17. Run Chart

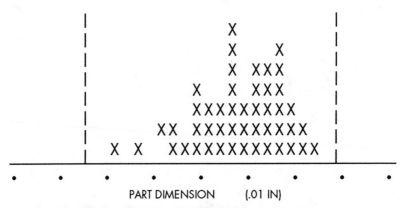

Figure 18. Distribution of part sizes, without benefit of automatic control system

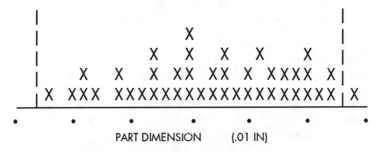

Figure 19. Distribution of part sizes when automatic control system is used

was placed under the control of an automatic controller. (Data courtesy of W. W. Scherkenbach of Ford Motor Company). The histogram tells at a glance that the automatic controller was overcontrolling. It produces a statistical distribution that is broader than before the control was used.

A histogram can reveal a great deal about a process. A "double humped" histogram, for example, is evidence that outputs from two different processes have been mixed.

A Seventh Tool: Scatter Charts
Figure 20 shows one variable plotted against another. The points form a pattern which suggests a correlation.

An Eighth Tool: Control Charts
Figure 21 shows a control chart for the production of a part. The upper and lower control limits indicate the limits of "normal" or "usual" production.

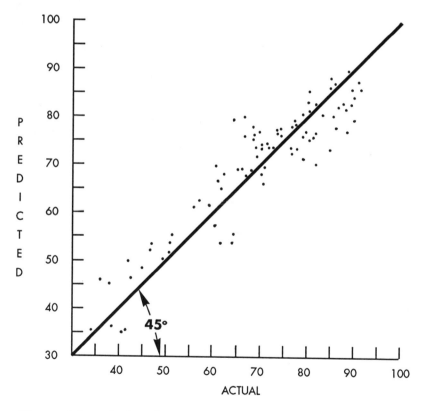

Figure 20. Relationship of Actual Rates of Registration to Predicted Rates (104 cities 1960)

A Vision of the Future

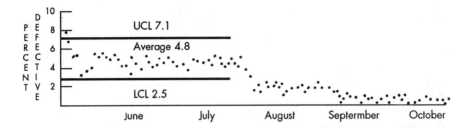

Figure 21. Deming, W.E.; Quality, Productivity, and Competitive Position; pg 178

As long as the data points fall inside these limits, the operator (and the manager) should leave the process alone. Every process exhibits such variation. The job of the management is to find out what are the normal limits associated with this variation and decide whether they need to be improved.

CONSTANT IMPROVEMENT IS THE NAME OF THE GAME
Earlier we commented that meeting specifications is not enough. Genichi Taguchi has presented an interesting example of why this is so in his article, "The Role of Metrological Control for Quality Control." In figure 22 we show the distributions of characteristics from television sets manufactured by Sony in the USA and Japan. These characteristics pertain to the color and control system. The vertical bars indicate the specification limits. As indicated in the diagram, all of the sets manufactured in the USA

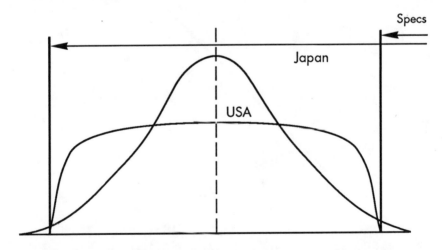

Figure 22. Distribution of Quality Characteristic for color control in television sets manufactured by Sony in USA and Japan.

met the specifications. They were within tolerances. The sets from Japan did not meet the specifications. Some fell outside the limits. Yet when customers were asked, it was found that they preferred the sets from Japan. Why should this be?

We can understand better why the Japanese made sets are preferred if we recognize that the vertical line at the center of the diagram represents the most desirable value. Any deviation from this value represents a loss. We do not know precisely how to describe this loss, but it makes sense to say that the loss increases as the distance from the most desired point. A simple "loss function" is a parabola, as sketched in figure 23.

If we multiply this loss function by the distributions, the area under the curve gives the weighted loss associated with the distributions. As shown in Figure 24, this loss is, on average higher for the U.S. produced sets, *even*

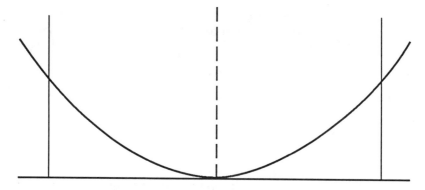

Figure 23. A plausible "loss function" indicates a loss in quality proportional to the square of the deviation from "best value".

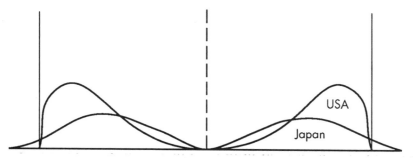

Figure 24. The area under the curve when the loss function is multiplied by the distribution function indicates the weighted loss associated with each distribution.

though all of the sets produced in the U.S. were within specifications and some of the Japanese sets were not!

There is a fallacy in setting tolerance limits. The difference between the performance of a set that falls just inside the limits is not very different from the performance of a set that falls just outside. Yet, one will be sold and the other will be reworked. We have been so accustomed to using "minimum acceptable" as a measure of "quality" that a deviation from this practice seems strange. As Dagwood Bumstead has said: "That makes a lot of sense if you don't think about it."

THE PROGRESSION TO EXCELLENCE

We do not have too much experience in observing how a company moves from near bankruptcy to excellence. In general, based on these experiences, it appears that the transition tends to follow these stages:

Stage 0: The management expresses concern only over market share, profits and return on investment.

Stage 1: The management is concerned about quality of the product because of impact on warranty cost and customer complaints, loss of market share. The action taken is to add more inspectors.

Stage 2: Management recognizes that control of the production process will lead to less waste and a lower cost to obtain acceptable products. QC is added to manufacturing.

Stage 3: The results of QC are limited by reactions of personnel so management begins to emphasize quality management. Manufacturing introduces statistical quality control.

Stage 4: Management asks that SQC and quality management methods be applied to all departments which border on the produciton department. (Purchasing, transportation, warehousing, etc.)

Stage 5: Management applies quality management principles to R&D, and to engineering. (There is considerable resistance because these departments have a hard time believing they have anything to do with quality problems)

Stage 6: Management recognizes that quality management principles will be useful if applied to all departments of the enterprise.

Stage 7: Management proclaims (and acts accordingly) that "CWQC is the company policy" specifically, this means:

    Quality is first priority
    Customer oriented decision criteria
    Personnel policies respect humanity
    All departments coordinate and cooperate
    All employees involved in improvement
    Good communication based on
        Factual data
        Statistical quality control
    Solid relations with suppliers

THE BARRIERS TO PROGRESS
The most important barrier is the image that most managers have of the enterprises they lead. Figure 25 shows both an organization chart and a flow chart.

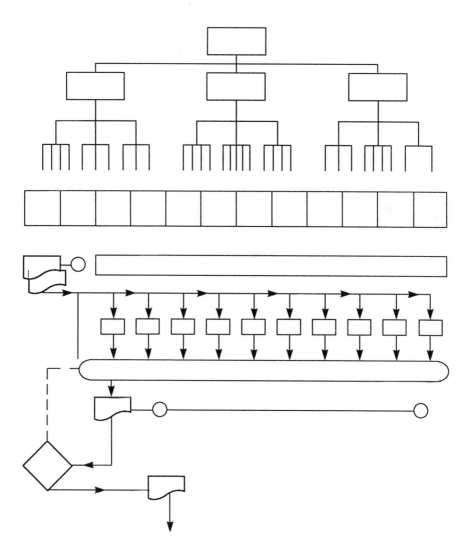

Figure 25. An organization chart and a flow chart are two different "maps" of the organization and its activities. It makes a difference which one comes to mind when the manager decides to take an action.

If the manager thinks of the organization in terms of the organization chart, and tries to improve the situation, it is likely that one set of actions will be taken. On the other hand, if the manager thinks in terms of flow charts, a different set of actions will come to mind. The organization man cannot see the flow processes for the tree diagram.

Tree Structures Versus Flow Processes
Where Attention Will be Focused

TREE	FLOW PROCESS
Motivate People	Remove Barriers
Find Out Who is Wrong	Find What is Wrong
Allocate Responsibility	Study the Process to Prevent Flaws
Fix Everyone's Attention on "The Bottom Line"	Fix Everyone's Attention on Quality
Call for Clear Measures of Productivity	Call for Clear Measures of Quality
Call for and Reward Individual Achievement	Call for and Reward Group Achievement
Give Crisp "Marching Orders" "Do Your Job"	Establish Well Defined Procedures "Help Me to Help You to Do Your Job"

THE SECOND PRINCIPLE IN QUALITY MANAGEMENT

A person who sees the organization in terms of a tree diagram will often try to improve the system by a technique such as "Management and Objectives". Of course it is a good idea to have objectives and to manage in such a way as to try to achieve them. It is also to discuss objectives with the people who report to you. For one thing, you need to know if their objectives are consistent with your own. Many people advocate going beyond that: They propose that each manager negotiate with the next level down and set targets for personal achievement or the achievement of an organization. Setting such targets removes from manager the responsibilty to improve the system. Observations of the results of such activities suggests the second principle, also known as the Perversity Principle:

The Second Principle in Quality Management (the perversity principle): *If you try to improve the performance of a system of people and machines by setting numerical goals and targets for their performance, the system will defeat you and you will pay a price where you did not expect it.*

This principle goes down hard with most audiences. Those who have learned to manage through flow charts and according to the first principle, understand it. Those who continue to see the world as though all organizations *operated* according to tree diagrams find the idea too strange to adopt.

CONCLUSION — A PARABLE

I am indebted to Lewis A. Rhodes of the Association for Supervision and Curriculum Development in Alexandria, Virginia for this little story.

Once upon a time there was a captain of a ship who carried cargo between San Francisco and Tokyo. He followed a straight line on the map, as shown below.

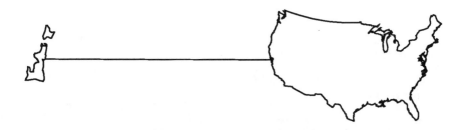

One day a passenger by the name of Deming came aboard and said, "Captain, why don't you follow a route like this?" and he drew a curved line as shown in the next figure.

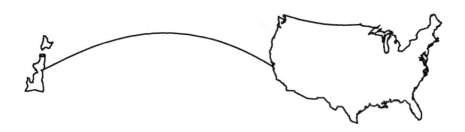

The Captain was not amused. He said, "Look here, I do not have time to follow such a route. I do not have the fuel. My customers are waiting. Everyone knows the shortest distance between two points is a straight line. I tell my men to keep the compass heading right on Tokyo. A straight line means a good bottom line."

Dr. Deming got off the boat in Tokyo and he began to teach the Japanese captains how to navigate. They followed the polar route. After a while the American captain noticed that his competitors were offering lower rates and faster service. He became quite agitated and when in Tokyo harbor he demanded to inspect the other ships. He found to his amazement that they had the same power plant, the same hull design, the same amount of cargo space. The only thing he noticed was that the crew seemed to be going about their work with a certain confidence. "That's it," he said, "it's cultural."

The one thing the Captain did not examine was the image of the world that was in the other Captain's head. He did not recognize that if you have a polar projection of the Earth you see things differently than if you have a Mercator projection.

Too many managers still operate from the premises of the flat earth society. The techniques are there to be used. They are simple, probably simpler than many of the methods now in use. They are easy to learn. All it takes is to abandon the idea that the Earth is flat.

References
Deming, W. Edwards. 1982. *Quality, Productivity and Competitive Position.* Center for Advanced Engineering Study. Massachusetts Institute of Technology. Cambridge, Massachusetts.
Feigenbaum, Armand V. 1983. *Total Quality Control, 3rd Ed.* McGraw Hill Book Company. New York.
Fuller, F. Timothy. 1985. There's just not much work in anything. Presented at the Second Ellis Ott Conference. New Brunswick, New Jersey.
Garvin, David A. Sept-Oct. 1983. Quality on the line. *Harvard Business Review,* pp. 64-65.
Georgescue-Roegen, Nicholas. 1971. *The Entropy Law and the Economic Process.* Harvard University Press. Cambridge, Massachusetts.
Harbour, James and Associates. September 1982. Presentation for "Quality Day". Jackson Community College. Jackson, Michigan.
Harbour, James and Associates. May 1983. *Automotive Industries Magazine.*
Hunter, Stuart J. 1983. "Theory Sigma", a privately circulated paper postulating that there exists in every place of work a vast quantity of information that is ignored and could be used to predict and prevent flaws but is treated as "noise" and not subjected to proper analysis.
Pirsig, Robert M. 1974. *Zen and the Art of Motorcycle Maintenance.* Morrow Publisher. New York.
Shewhart, Walter. 1931. *Economic Control of Quality in Manufactured Product.* D. Van Nostrand Company, Inc. New York.
Stueart, Robert D. and Eastlick, John T. 1981. *Library Management.* Libraries Unlimited, Inc. Littleton, Colorado.
Taguchi, Genichi. "The Role of Metrological Control for Quality Control" (privately released). Available from the American Supplier Institute, Inc. Romulus, Michigan.
Takamiya, Makoto. September 1979. Japanese Multinationals in Europe: Internal Operations and Their Public Policy Implications. *Discussion Paper Series.* International Institute of Management. Wissenschaftszentrum, Berlin.
Taylor, F. W. 1911. *The Principles of Scientific Management.* Harper and Brother. New York and London.
Tribus, Myron. 1961. *Thermostatics and Thermodynamics, An Introduction to Energy, Information and States of Matter.* D. Van Nostrand Company. Princeton, New Jersey.

About the Editors

Victor R. Dingus, P.E., C.Q.E., is a technical associate in the Textiles Fibers Division, Tennessee Eastman Company. He has broad experience in inventory and production management, equipment justification, organizational design, and material handling systems. He has coordinated a corporate value engineering program, and developed and implemented a corporate productivity and quality improvement strategy. He currently provides leadership and consulting in design and implementing performance systems to accelerate quality management throughout the company. Dingus is a part-time faculty member at East Tennessee State University. He is a senior member of IIE and senior member of ASQC. Dingus holds BSIE and MBA degrees from the University of Tennessee, Knoxville.

William A. Golomski, P.E. is president, W. A. Golomski & Associates, international technical and management consultants. He is sought after as a speaker for management meetings worldwide. His consulting experience includes dealing with very small to multi-billion dollar firms. Golomski is one of the judges for the Malcolm Baldridge National Quality Award program. He is past director of IIE's Quality Control and Reliability Engineering Division.